High vacuum techniques for chemical syntheses and measurements

High vacuum techniques for chemical syntheses and measurements

P. H. Plesch, MA, PHD, SCD, CCHEM., FRSC
Emeritus Professor of Physical Chemistry
[University of Keele]

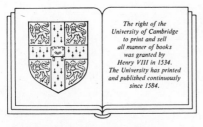

The right of the
University of Cambridge
to print and sell
all manner of books
was granted by
Henry VIII in 1534.
The University has printed
and published continuously
since 1584.

CAMBRIDGE UNIVERSITY PRESS

CAMBRIDGE

NEW YORK   PORT CHESTER

MELBOURNE   SYDNEY

Published by the Press Syndicate of the University of Cambridge
The Pitt Building, Trumpington Street, Cambridge CB2 1RP
32 East 57th Street, New York NY 10022, USA
10 Stamford Road, Oakleigh, Melbourne 3166, Australia

First published 1989

Printed in Great Britain by the University Press, Cambridge

British Library cataloguing in publication data
Plesch, P. H.
High vacuum techniques for chemical
syntheses and measurements.
1. Organic compounds. Synthesis. Vacuum technology
I. Title
547.2′028

Library of Congress cataloguing in publication data
Plesch, P. H.
High vacuum techniques for chemical syntheses and measurements
P. H. Plesch.
   p. cm.
Includes bibliographies
ISBN 0 521 25756 5
1. Vacuum. 2. Chemistry—Manipulation. I. Title.
QD63.V33P53 1989
541.3—dc19 88–20731 CIP

ISBN 0 521 25756 5

Dedication

This book is lovingly dedicated to my ever-helpful wife
Traudi Plesch, MBE, to whom I owe so much.

Contents

Preface

The first words of this book are devoted to my friend Dr Steve Pask, who had urged me to undertake its writing in collaboration with him, which I eventually did. His contributions include much of the text and some valuable constructive criticism, and it was to the great regret of both of us that circumstances forced him to withdraw from the project. Without him this book would never have come to be. However, I alone take responsibility for what it is now.

Amongst my hitherto unacknowledged debts is one to H. A. Skinner, my PhD supervisor (under the general direction of Michael Polanyi), who introduced me to, amongst many other things, the arts of glass blowing and vacuum technique when I started at Manchester in 1944. He, in turn, had learnt them from H. W. Thompson at Oxford.

In the context of practical, especially vacuum line, chemistry, it is my pleasure to express a warm appreciation of Fred Fairbrother, a big man in every way. He had a reputation of sequestering scarce apparatus in his laboratory through a 'non-return valve', but towards me he always showed the greatest generosity in sharing equipment and his incomparable store of practical experience and theoretical insight.

It would be misleading, when writing about the Manchester University Chemistry Department in the late 1940s, to omit the impact made upon our little community by that human ball-lightning, Michal Szwarc. His influence was pervasive and stimulating, not least because so many of us were devising new devices to do new things. His background in chemical engineering made him a most valuable ally in our contant fight against the 'malice of the inanimate' (*die Tuecke des Objekts*).

Last, I must acknowledge my very great debt to M. G. Evans who had succeeded Michael Polanyi in the Chair of Physical Chemistry when I moved from Manchester to Keele in January 1951. When I asked whether the Chemistry Department at the newly founded University College of North Staffordshire (later to become the University of Keele) could buy a few items of equipment from my research laboratory, he encouraged me to take away the whole content of my laboratory (which I did, in 24 tea-chests) and 'never mind any payment'. It is difficult to exaggerate the importance which this generous and understanding gesture had for the rapid establishment of my Polymer Research Group at Keele.

The techniques which my many collaborators helped me to evolve have proved their worth, and have been imitated and improved in laboratories all over the world. Inevitably, this book contains a preponderance of Keele devices and I acknowledge here my debt to all my co-workers, and to our ingenious and persistent glass blower, Mr C. C. Cork, who contributed to the stock. I owe a special thanks to my friends Drs R. O. Colclough and R. N. Young whose perspicacious comments have helped me to improve the original script.

Of course, I have done my best to assemble from the literature and through personal contacts as many as possible of the most useful devices and techniques developed elsewhere. Since I am hoping that this book will see a second edition (at least!), I encourage anyone who cannot find his favourite apparatus here to write to me with a drawing, a description and a reference, and the same goes for anyone who knows of a better version of something that is included. In this way the second edition should become a markedly more useful product.

Finally, I thank the Cambridge University Press for their forbearance when the production of this book was delayed excessively by the multiple upheavals accompanying my retirement.

Professor Emeritus, University of Keele,   P. H. Plesch
January, 1988

Acknowledgements

Acknowledgements of permission to reproduce figures
The exact references to the publications whence the figures have been taken are given in the text.

American Chemical Society (Copyright)
Figs. 3.3, 5.8

Chemistry and Industry (London)
Figs. 3.21, 3.27

La Chimica e l'Industria (Milano)
Fig. 3.15

European Polymer Journal
Fig. 3.25

GIT Verlag, Darmstadt
Figs. 2.13, 3.4, 3.9, 4.3, 5.7

Huethig und Wepf Verlag, Basel
Figs. 3.13, 3.16, 3.18, 4.5, 4.7

International Laboratory
Fig. 3.12

Journal of Macromolecular Science
Fig. 5.4

Dr B. Krummenacher
Figs. 2.16, 3.6, 3.10, 3.24

The Royal Society of Chemistry
Figs. 2.10, 3.17, 3.22, 3.23, 3.26, 4.4, 5.3, 5.5, 5.11

Société Chimique de France
Fig. 5.2

J. Wiley and Sons, Inc.
Fig. 3.8

Introduction

I.1. The purpose and scope of this book

The purpose of this volume is to enable a chemist to assess the possible utility of high vacuum technique for his purposes, to design and build a high vacuum system, and to do good chemistry with it. This book is not a textbook of vacuum theory, glass blowing, sophisticated analytical techniques, or anything else; it also does not deal with the techniques of gas reactions. It is not a replacement for 'Green', 'Shriver', or 'Yarwood' (see below). There is much in these, and of course in the large specialist volumes, which is not in this one.

In other words, the present book does not aim at being comprehensive in any respect, but is the best compromise the author could achieve between necessary theory and sufficient useful advice. However, here can be found numerous practical, traditional tricks and hints not found elsewhere, because the aim has been to produce a guide and companion for the worker in the laboratory.

Lest it be thought that the techniques to be described are only for the affluent institution, the author wishes to emphasise that he has only ever worked in under-funded laboratories, was brought up before, during, and after the war in a 'string-and-sealing wax' tradition, and ever heeded Lord Rutherford's exhortation to his team at the Cavendish Laboratory: 'We've got no money, boys, so we've got to think.' Very many of the author's collaborators came from or went to laboratories far better endowed than his, but were all the better for the lessons learnt in a 'do it yourself' atmosphere. That is also the reason why there are few references to sophisticated, expensive apparatus and why this book may have a slightly old-fashioned look.

I.2. Usages

I.2.1. References
The references given are simply those which are deemed most useful, and there is no pretence that they are comprehensive. The devices quoted are not the earliest nor the most famous, but generally those judged to be of the greatest use and simplicity.

I.2.2. Figures

Most of the very numerous figures are to be understood as working sketches rather than scale drawings. They should, however, be completely adequate for a glass blower to make the pieces concerned.

Invitation: If anyone is unable to make any of the pieces shown here, and cannot arrange to have it made, he should write to The Laboratory Superintendent, Chemistry Department, University of Keele, Staffordshire, ST5 5BG, England, indicating exactly what is wanted. He will then be sent a quotation for the cost of having the piece made by the Departmental glass blower.

I.2.3. Units

Throughout this work SI units have been used, with the exception of pressure, which is given in Torr. For formulae, table headings etc. the Quantity Calculus has been used, as advised by M. L. McGlashan, '*Physicochemical Quantities and Units*', Royal Society of Chemistry Sales and Promotions Department, Burlington House, London, W1V OBN.

I.3. Books

There are astonishingly few books which deal with the chemist's vacuum manipulations (but many for the physicist), and even those marginally connected with the subject are not numerous. Each of the books named below contains some matters of interest in the present context, but the list is certainly not comprehensive, and if a volume which is considered of value has been omitted, the author asks that it be brought to his notice.

The books which are considered useful have been listed here in alphabetical order of their authors:

R. Barbour, *Glassblowing for Laboratory Technicians*, Pergamon Press, Oxford, 1968.

L. Bretherick, Ed., *Hazards in the Chemical Laboratory*, 3rd Edition, Royal Society of Chemistry, Burlington House, Piccadilly, London, W1V OBN.

S. Dushman, *Scientific Foundations of Vacuum Technique*, 2nd Edition, Wiley, New York, 1962.

G. W. Green, *The Design and Construction of Small Vacuum Systems*, Chapman and Hall, London, 1968.

H. W. Melville and B. G. Gowenlock, *Experimental Methods in Gas Reactions*, MacMillan, London, 1964.

L. M. Parr and C. A. Hendley, *Laboratory Glassblowing*, G. Newnes Ltd., London, 1956.

A. J. B. Robertson, D. J. Fabian, A. J. Crocker and J. Dewing, *Laboratory Glassblowing for Scientists*, Butterworth Scientific Publications, London, 1957.

D. F. Shriver and M. A. Drezdzon, *The Manipulation of Air-Sensitive Compounds*, 2nd Edition, Wiley-Interscience, New York, 1986.
M. Wutz, H. Adam and W. Walcher, *Theorie und Praxis der Vakuumtechnik*, Viehweg und Sohn, Braunschweig (FRG), 1986.
J. Yarwood, *High Vacuum Technique*, Science Paperbacks and Chapman and Hall, London, 1975.

I.4. Articles

Apart from the numerous papers and articles quoted in the book itself, there are a few general ones, quoted below, which contain useful material for the purposes under consideration here. One of the earliest, and still amongst the most useful, is:

L. J. Fetters, Procedures for Homogeneous Anionic Polymerisation, *J. Res. Natl. Bureau of Standards, A. Phys and Chem*, **70A**, 421 (1966).

Some useful modern tricks can be found in:

A. L. Wayda and J. L. Dye, A Versatile System for Vacuum-line Manipulations, *J. Chem. Ed.*, **62**, 356 (1985).

Although the next item is mainly concerned with non-vacuum methods, it is worth a close look:

G. B. Gill and D. A. Whiting, Guidelines for Handling Air-Sensitive Compounds, *Aldrichimica Acta*, **19**, 31 (1986).

The following paper, which is very clear and concise, has been extensively quoted because it is not readily available in the English-speaking world:

O. Nuyken, S. Kipnich and S. D. Pask, Ueber die Manipulation empfindlicher Substanzen im Hochvakuum, *GIT Fachz. f.d. Laboratorium*, **25**, 461 (1981).

1 Fundamentals

1.1. Reasons for using high vacuum techniques

1.1.1. Introduction

High vacuum technique (h.v.t.) is one of several types of experimental technique which can be employed to obtain a controlled experimental environment. The most usual reason for wanting this is the necessity to exclude oxygen and/or water and, less commonly, carbon dioxide from the reaction being studied. Perhaps the most primitive example of creating a controlled environment is the use of a 'soda-lime' tube to protect a store of sodium hydroxide from the ingress of carbon dioxide.

Before considering the advantages and disadvantages of h.v.t. as compared with some of the alternatives, it is important to point out that, as with all experimental chemistry, the first questions to be answered are what information is required from the planned experiment, what level of impurities can be tolerated, and what resources are available. It is worth remembering the first two Laws of Experimentation:

(1) What one gains in rigour one loses in flexibility.
(2) The greater the rigour of the technique, the longer the experiment takes (the perfect experiment takes an infinite time).

Having decided that a controlled environment is needed to exclude certain normal atmospheric components, one must then decide on the levels to which concentrations of unwanted compounds must be reduced and below which they must be maintained.

There is a sequence of experimental techniques ranging from the 'open beaker on the bench' to the extremely rigorous h.v.t. used for studying the kinetics of radiation-induced cationic polymerisations. Each technique in the series has its characteristic level of extraneous materials, and as the level of 'cleanliness' is raised, there is a concurrent loss of flexibility.

If a controlled environment is necessary for the desired experiment, it is useful to examine some of the techniques alternative to h.v.t., to estimate the level of impurities involved and to assess their advantages and disadvantages as compared to h.v.t.

When contemplating the choice of technique one must keep in mind two further, highly relevant, Laws of Experimentation:

(3) Murphy's Law: If anything can go wrong, it will go wrong. This is a very strong inducement to keep every set-up as simple as possible.
(4) Cheops' Law: Every construction takes longer and costs more than the most pessimistic estimates.

1.1.2. The inert gas blanket
The use of a large funnel attached to a stream of inert gas situated above the bench is particularly useful when opening bottles of hygroscopic or hydrolytically unstable materials. This technique relies for its success on a fast stream of 'clean' and 'dry' inert gas. It is of course wasteful when used over long periods, but it does allow almost the same flexibility of manipulation as on the open bench.

1.1.3. Dry-bag and dry-box
The move from the open bench to a closed system involves loss of flexibility, and so a greater degree of pre-experimental planning is necessary. With careful operation, the moisture content of the atmosphere in a dry-box can be reduced to such a low level that one can do experiments in which 10^{-3} mol l^{-1} of water in reagents is tolerable. To achieve this level it is important that the techniques be used intelligently, and it is worth examining

some of the critical steps, since even when the main part of an experimental programme is to be carried out by h.v.t., it is often expedient to use a dry-box or a dry-bag for some of the preparatory stages.

Both the dry-box and the dry-bag offer a considerable degree of experimental flexibility and are very suitable for larger scale preparative work in which catalytic amounts of impurities are unimportant. However, in this context it is very important to remember that some materials, and expecially natural products such as cork, wood and paper, have a high water content, however carefully they may have been dried and/or degassed, and that they should therefore never be used inside a dry-bag or dry-box.

A *dry-bag* is a bag, usually about 0.5×1 m, made of a transparent, inert and tough film or sheet which has an easily sealable entry port for inserting apparatus and several ports for admitting purging gas, electrical leads, etc., and at least two gloves so that operations can be done inside it.

The dry-bag is usually operated with a slight overpressure of the inert gas to hinder diffusion of unwanted gases into the bag. Simply running a stream of inert gas through the bag is not an efficient method of producing an inert atmosphere because the air in the lateral excrescences may take a long time to be swept out. In order to produce the inert atmosphere efficiently, the bag must be emptied and filled several times before starting the experiment. In order to find out the number of evacuation/filling cycles required to produce the desired 'controlled' environment, consider a bag having a volume of 50 l. Suppose the flattened bag has a residual volume of 1 l and the bag is then filled with 'pure' nitrogen. Since air contains some 20 % of O_2, after one evacuation/filling cycle the gas inside the bag will contain some 0.2 l of O_2 (*ca.* 9×10^{-3} mol at s.t.p.) and therefore the bag contains 0.4% of O_2. A second cycle will reduce the amount of O_2 to 4×10^{-3} l (*ca.* 2×10^{-4} mol at s.t.p.) and the gas in the bag now contains only 8×10^{-3}% of O_2. Of course, in the above calculation it has been assumed that the inert gas itself is 'pure', i.e. that its content of noxious gases is no greater than that which has been deemed tolerable for the purpose in hand, but it does indicate the level of atmospheric 'cleanliness' that can be obtained under optimum conditions. In this and in many other contexts one must remember that in relation to evacuation one is rarely, if ever, dealing with a true equilibrium, and that when the quantities of, for example, water adsorbed on glass or absorbed in organic materials (see p. 40), or volatile organic substances ad- or absorbed by rubber and plastics and greases may be of importance, only an all-glass, glass and metal, or all-metal system will be satisfactory.

The *dry-box*, also known as *glove-box*, consists essentially of a metal box with a viewing window and (at least) two rubber or plastic gloves, a port, usually with a 'lock' for the introduction of reagents or pieces of equipment, and ports for the in-flow and out-flow of inert gas, for electrical leads, etc. The atmosphere inside the box is established by a stream of well-dried and deoxygenated inert gas which is usually kept slightly above atmospheric pressure. The dry-box, as opposed to the dry-bag, is more capacious, permanent and robust. An additional advantage is that it can be made as large as necessary. For exceptionally well-controlled experimental conditions, and additionally as a means of reducing the amount of inert gas required, the inert gas can be recirculated via an appropriate purification train. This technique has the advantage that impurities in the inert gas have less opportunity of collecting within the aparatus.

1.1.4. Vacuum lines

For smaller scale operations or when catalytic amounts of impurities are important, a vacuum line is almost always more suitable and efficient. Although it takes longer to become familiar with a vacuum apparatus, once a certain degree of competence and confidence has been achieved, there is little to choose in terms of experimental time between h.v.t. and the rigorous use of a dry-box. The choice is often made simply on the basis of laboratory tradition (i.e. inertia) and know-how.

As a reason for not using h.v.t. it is often stated that its use leads to experimental results that cannot be reproduced on an industrial scale. *This is untrue*. A closed system, such as an all-glass vacuum line, has more in common with an industrial plant than the typical apparatus used at the laboratory bench. Furthermore, because of the considerably more favourable surface to volume ratios in a large plant, the typical concentrations of those impurities which originate from surfaces are more accurately reproduced by h.v.t. experiments than by the typical bench experiment. This is often reflected in the problems encountered during development work when bench experiments are being scaled up to pilot plant and beyond.

A decision to use a vacuum line does not necessarily imply that what is here called h.v.t. must be used. The vacuum lines familiar to many inorganic chemists, and employed for many radical polymerisations differ from the all-glass high vacuum lines which are the central theme of this book, but the distinction is not rigorous. Normally, the simpler lines are used for preparative experiments in which the exclusion of oxygen and not water is the major problem. As the calculation given earlier in this section shows, if the experimental volume is evacuated between repeated fillings with an inert gas, the amount of residual gaseous impurities rapidly approaches zero.

1.1.5. Summary

The advantages of using a vacuum rather than an inert gas environment are:

(1) Reproducibility of the experimental conditions is fairly easy to obtain. However, reproducibility of results in one set of circumstances is, by itself, no guarantee of their validity, since, especially with a vacuum line, the impurity effects can often be reproduced extremely well.

(2) With care, which usually means longer pumping times and greater rigour in the preparation of the experiments, the level of the impurities, such as water in liquid reagents or solvents, can be reduced to below 10^{-5} mol l^{-1}.

(3) Because of the very low concentrations of water and oxygen that can be achieved, and particularly because the systems are inherently closed, it is possible to use extremely strong drying agents such as $K + Na$ or $Pb + Na$ alloys, or hot Na vapour for purifying solvents and reagents without any considerable hazards.

(4) A particular advantage of using vacuum techniques is that liquid nitrogen can be used for the rapid freezing of materials and trapping of vapours without any danger that oxygen will be condensed out.

(5) In contrast, the single major disadvantage of h.v.t. is that it is generally limited to work involving small volumes of material (up to 500 ml, say), although solvent reservoirs up to a few litres are often used, and the firms making near-monodisperse polymers by anionic polymerisation do much of the preparative work with very large high vacuum systems.

(6) It is worth repeating that the idea that h.v.t. is inherently more time-consuming than dry-box techniques is a myth; with careful experimental planning, a greater degree of reproducibility and a lower level of impurities can be achieved by h.v.t. in a similar, if not shorter, time than that required for dry-box work. It is, however, true that it takes longer to master h.v.t. than to become familar with dry-box techniques.

(7) Finally, there are many operations, both preparative and investigative which *can* be done *without* h.v.t., but which are done *more efficiently*, *swiftly*, and *safely* by means of it.

1.2. Theoretical considerations

1.2.1. What is a vacuum?

The word vacuum in the English language is the same as the Latin word meaning a completely empty space. In practice, the word vacuum is used to describe any space in which the pressure is less than atmospheric pressure. Further, in order to specify the quality of a vacuum, various adjectives are used. Table 1.1. lists a selection of these terms, the associated pressure range, and the number of molecules per cubic centimetre that such a vacuum contains. It is clear that even the best known vacuum, in outer space, is not absolute. A knowledge of the amount of gas in a reaction volume at the operating pressure allows an assessment to be made of its possible effect on the reaction taking place.

Table 1.1. *Quality and nomenclature of vacua*

Pressure p/Torr	Number of molecules/cm³ at 273 K	Quality of vacuum
760	2.73×10^{19}	
100	3.5×10^{18}	Coarse
~ 20	—	Water-pump vacuum
1	3.5×10^{16}	
10^{-3}	3.5×10^{13}	Fine
10^{-6}	3.5×10^{10}	High (usually the chemist's best)
10^{-10}	3.5×10^{6}	Ultra-high
10^{-16}	4	Outer space

Once a high vacuum system has actually been used for 'chemistry', i.e. has contained compounds other than the constituents of air, it is very difficult and tedious to evacuate it to below 10^{-5} Torr.

1.2.2. Units
The practical chemist working with vacuum systems has, in the past, used practical units, such as millimetres of mercury and atmospheres, for measuring the quality of the vacua which he has produced. However, for the modern chemist it is important to have a coherent system of units in which no numerical factors are inherent. Within the SI system, the unit of pressure is the pascal, which has the units and dimensions given below:

$$Pa = m^{-1} \ kg \ s^{-1}$$

Since many measuring devices and indeed many chemists still work with other units, it is useful to have the relations between all the units in common usage:

$$1 \ atm = 760 \ Torr = 760 \ mmHg = 1.01325 \ bar = 1.01325 \times 10^{5} \ Pa$$

In this work all pressures will be given in Torr.

1.2.3. General structure of a high vacuum system
Although the structure and components of high vacuum systems (h.v.s.) will be discussed in detail in Chapter 2, it is essential for an understanding of what follows to give here an outline of what is involved (see Fig. 2.1.).

Almost all the high vacuum systems with which we will be concerned have at one end a 'rough' pump, usually a rotary oil pump, capable of attaining *ca.* 10^{-3} Torr. This is followed by a high vacuum pump which can attain *ca.* 10^{-6} Torr, which is followed by a cold trap, the purpose of which is to condense out any volatile matter to prevent it entering, and possibly damaging, the pumps; and beyond this is the vacuum trunk line to which are attached the various reservoirs, measuring devices, reactors, etc., the evacuation of which is the purpose of the whole installation. Usually each of the components is separated from its neighbours by a tap or valve. Also, it is usually expedient to have a direct connection between the rough pump and the vacuum line so that this can be brought down to *ca.* 10^{-3} Torr before the high vacuum pump is engaged; this process is called 'roughing'.

1.2.4. Theory of pumping

In order to be able to design even the most elementary vacuum line, it is necessary to know something of the basic theory concerning the movement of molecules within the system. No attempt will be made here to instruct the reader in the details of vacuum physics and therefore the formulae given below have been kept simple. A more detailed discussion of the theory of gaseous flow can be found in Dushman's excellent review (Dushman, 1962). The object of the following discussion is simply to allow the reader to assess, without too much effort, the approximate efficiency of the system he is planning.

A chemist's vacuum line will often be pumped out several times within any working day and therefore the time taken to reach a useful vacuum should be as short as possible. The time taken by a system to reach the described vacuum depends upon the volume of the system, its internal surface area, the materials used for its construction, its cleanness, the speed of the pumps and the aerodynamic conductance of the system.

The conductance of any system depends on the nature of the gas flow; at higher pressures a viscous flow regime prevails and at lower pressures ($p < 10^{-1}$ Torr) a molecular flow regime. Turbulent flow, seldom considered when assessing the efficiency of a vacuum system, is encountered only when the pressure in the system is close to atmospheric pressure. As a general rule, if the efficiency of the system is adequate for the viscous flow regime then it will also be suitable for turbulent flow.

The change from a viscous to a molecular flow regime occurs when the mean free path $\bar{L}$ of the gas molecules in the system exceeds the minimum physical dimensions of the system. The mean free path is a measure of the average distance a molecule travels between collisions. The derivation of $\bar{L}$ involves a number of assumptions about the ideality of the gas and the nature of the collisions and by definition some 63.2% of the molecules in a particular gas collide with other molecules within the distance $\bar{L}$. The mean free path for any gas can be calculated from Equation (1.1)

$$\bar{L} = \hat{k}T/2^{\frac{1}{2}}d^2p \tag{1.1}$$

where $\hat{k}$ is Boltzmann's constant, T the absolute temperature, d the effective diameter of the molecule, i.e. the distance between the centres of the two molecules at collision, and p is the pressure.

Since the main pumping lines are usually constructed from tubing with an internal diameter (i.d.) of *ca.* 2.5 cm, viscous flow predominates until the pressure is reduced to *ca.* 10^{-1} Torr, when the high vacuum pump is usually switched into the system.

The conductance C of a duct for a flow of gas is the rate of flow under unit pressure gradient. The conductance C_v of a system in which *viscous* flow predominates can be calculated from Equation (1.2):

$$C_v = \pi r^4 \bar{p}/8\eta l \tag{1.2}$$

where r is the radius of the tube, $\bar{p}$ is the average pressure between the two ends of the tube being considered, η is the viscosity of the gas, and l is the length of the tube.

From Equation (1.2), which is derived from the Poiseuille formula for viscous flow, it can be seen that C_v is directly proportional to $\bar{p}$. For air at 25 °C, Equation (1.2) becomes Equation (1.3):

$$C_v/(\mathrm{m^3\,s^{-1}}) = [2.16 \times 10^4 (r/\mathrm{m})^4 (\bar{p}/\mathrm{Pa})]/(l/\mathrm{m}) \tag{1.3}$$

When making rough estimates of C_v for a system, it is always best to take the lowest value for $\bar{p}$ (i.e. simply the final pressure required), so that one obtains the most conservative estimate of the overall conductance. With a knowledge of the conductance of the system, it is possible to derive the net pumping speed from Equation (1.4):

$$S_{\mathrm{net}} = S_p C/(S_p + C) \tag{1.4}$$

where S_p is the nominal speed $(\mathrm{m^3\,s^{-1}})$ of the pump and C the conductance of the system. As this equation is also valid for the molecular flow regime, the subscript v for C has been omitted.

The time t_c which a system takes to achieve a pressure (the crossover pressure p_c) at which the high vacuum pump can be switched into the system is given by Equation (1.5):

$$t_c = (V/S_{\mathrm{net}}) \ln (p_a/p_c), \tag{1.5}$$

where p_a is the starting pressure and V the volume of the system. The crossover pressure is usually *ca.* 10^{-1} Torr, and if one assumes that $p_a = 760$ Torr, Equation (1.5) can be approximated by Equation (1.6):

$$t_c/\mathrm{s} = 9(V/\mathrm{m^3})/(S_{\mathrm{net}}/\mathrm{m^3\,s^{-1}}) \tag{1.6}$$

An efficient roughing line will have a net pumping speed as nearly equal to the normal pumping speed as possible, so that the roughing time t_c is the shortest possible with the pump available.

As an example, consider a system with a volume of 0.1 m³ which consists of a tube 1 m long and a flask at the end remote from the pumps. Further, assume that the available pump has a nominal pumping speed of $S_p = 8 \times 10^{-3}\,\mathrm{m^3\,s^{-1}}$. From the above equations we can now calculate how

Table 1.2. *The effect of the size of a duct on the pumping characteristics*

Radius r/m	Conductance $C_v/m^3\,s^{-1}$	$S_{net}/m^3\,s^{-1}$	t_c/s
1.0×10^{-2}	2.16×10^{-3}	1.7×10^{-3}	5.42×10^2
1.25×10^{-2}	5.27×10^{-3}	3.18×10^{-3}	2.90×10^2

long it will take to reach a pressure of 10^{-1} Torr at which point the high vacuum pump can be switched into the system. In order to emphasise the effect of a change in the diameter of the tube on the pumping time, we consider (*a*) a tube with $r = 1$ cm and (*b*) a tube with $r = 1.25$ cm. The result is set out in Table 1.2. It can be seen that increasing the diameter of the tubing by 25 % reduces the pump-down time by almost 50 %.

The time taken for a system to reach a desired pressure within the molecular flow regime cannot be calculated as simply as it can for viscous flow, because the outgassing of the various materials used in the construction of the system starts to play an important role in terms of the quantities of gas which have to be removed. However, since the product of the pressure in the system and the net pumping speed is equal to the total gas load (measured in units of pressure × volume) it is clear that the net pumping speed (Equation (1.4)) should be kept as large as possible. The molecular flow conductance of a long tube is given by

$$C_{mol} = 2(8\pi \mathbf{R}T/M)^{\frac{1}{2}}r^3/3l \qquad (1.7)$$

where $\mathbf{R}$ is the gas constant and M the molar mass of the gas being pumped. For air at 25 °C Equation (1.7) reduces to Equation (1.8):

$$C_{mol}/m^3\,s^{-1} = 9.8 \times 10^2(r^3/m^3)/(l/m) \qquad (1.8)$$

It is important to note that in the molecular flow regime the conductance depends on the *cube of the diameter* of the tube and the $-\frac{1}{2}$ power of the molecular weight of the gas being pumped, but it is independent of the pressure.

Equation (1.7) is the general equation for any individual component of the system under molecular flow conditions and with it one can calculate the overall pumping time for a system, if the conductances of the individual components of the system are known. The effective pumping speed S_e at any point along the vacuum line is given by:

$$1/S_e = 1/C_e + 1/S_p \qquad (1.9)$$

and C_e, the effective conductance of the system from the pump to the point of interest is given by:

$$1/C_e = 1/C_1 + 1/C_2 + 1/C_3 \ldots \qquad (1.10)$$

where the subscripts 1, 2, 3,... refer to each of the individual components which are in series in the system up to the point of interest.

An important contribution to the conductance of any individual component in the system is due to the conductance C_0 of its orifice, which is given by Equation (1.11):

$$C_0 = (\mathbf{R}T/2M)^{\frac{1}{2}}r^2 \qquad (1.11)$$

in which r is the radius of the orifice. Although this is strictly valid only if the pressures on both sides of the orifice produce mean free paths which are at least ten times as great as the orifice, it can actually be used over a much wider range of pressures because the resulting errors are not serious for the present type of calculation. For air at 25 °C Equation (1.11) becomes:

$$C_0/\mathrm{m^3\,s^{-1}} = 3.6 \times 10^2 \, (r/\mathrm{m})^2 \qquad (1.12)$$

Any component can be considered as a combination of a tube of a given length and diameter and an orifice, the latter becoming more important to the overall conductance of the unit as the length of the component decreases. Consider a glass tap with a length of 3 cm and a bore of i.d. 1 cm. From the equations given above one can calculate that the conductance (for air at 25 °C) is equal to 2.82×10^{-3} m^3 s^{-1}; for a tap having a length of 2.5 cm and a bore of 0.8 cm the conductance is 1.75×10^{-3} m^3 s^{-1}. Therefore a decrease in the conductance of ca. 40 % resulted from a decrease in the bore of only 20 %. This shows why one should always use the largest available tap. Another interesting aspect of such calculations is the question of how to include the bends. Although an accurate calculation would need the shape of the bend (a sharp bend will have a lower conductance than one of large radius) a simple approach is to consider that at each bend the tube is interrupted. For each bend the conductance calculation will thus include two orifices of the same diameter as the tube from which the bend is constructed. Thus, the tube is treated as having an effective length greater than its actual length, the notional increase in length depending on the number of bends.

Lastly, the cold traps (see Section 2.2.) make an important (negative) contribution to the conductance of any vacuum system.

It should be remembered that the above treatment gives only approximate estimates of the pumping speeds and conductances at various points along the vacuum line. If, however, each component of the planned vacuum line is optimised with the help of the above equations, it will certainly be the best available within the prevailing limitations.

1.2.5. Some physico-chemical features peculiar to closed systems

1.2.5.1. General introduction.
The physical chemistry of closed systems, such as an evacuated line isolated from the pumps, is of course implicit in the general physical chemistry which should be familiar to the readers of

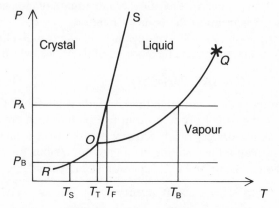

Fig. 1.1. Idealised p–T phase diagram, showing triple point O and critical point Q at temperature T_T, freezing temperature T_F and boiling point T_B at pressure p_A, and sublimation temperature T_S at pressure p_B.

this book. However, such systems, and especially the phase relationships prevailing in them, receive little emphasis in general courses of chemistry and that is the reason for pointing out here some of their characteristics and the ways in which they differ from open systems in which the pressure variable is fixed by the atmosphere.

The phase relationships for a normal substance are represented in the phase diagram shown in Fig. 1.1, which is interpreted in terms of the Phase Rule

$$P + F = C + 2 \qquad (1.13)$$

where P = number of Phases coexisting at equilibrium, C = number of Components, F = number of Degrees of Freedom. Since we are dealing with a single compound, $C = 1$, hence $F = 3 - P$. The boiling point is the temperature at which liquid and vapour are in equilibrium, i.e. $P = 2$ and so $F = 1$. Under 'open' conditions, the pressure is fixed by the prevailing atmospheric pressure (p_A in the diagram) and so the one available degree of freedom has been fixed and therefore so has the boiling temperature, T_B, by the intersection of the line $p = p_A$ with the vapour pressure curve OQ. In contrast to this, in the closed system the condition $F = 1$ means that either p or T can be fixed arbitrarily between the limits set by O and Q, so that the liquid has no 'boiling point', and is said to be 'under its own vapour pressure'. The point Q is the critical point beyond which the distinction between liquid and vapour, and also the meniscus of the liquid, disappears.

The normal freezing point of the liquid under pressure p_A is given by T_F, and OS is the melting curve of the substance, i.e. the locus of the points defining the co-existence of solid and liquid. If we measure the freezing point of a liquid in a closed system, the Phase Rule tells us that since at that temperature all three phases will be in equilibrium, $F = 0$, and we obtain the temperature T_T of the triple point, O. This distinction between the normal freezing temperature and the triple point temperature, although usually only a few tenths of a degree, can be important when one is using the freezing points of a series of pure liquids to calibrate a thermocouple or other type of thermometer in a vacuum system.

The phase diagram also illustrates why some substances which melt at normal pressure, will sublime at a lower pressure: the line $p = p_B$ intersects at T_S the locus OR of the points defining the solid–vapour equilibrium, i.e. at the pressure p_B the substance will sublime at the temperature T_S. Sometimes the opposite behaviour is observed, namely that a substance which sublimes at normal pressure will melt in a vacuum system 'under its own vapour pressure'. This is a non-equilibrium phenomenon and occurs if the substance is heated so rapidly that its vapour pressure rises above that of the triple point; this happens quite frequently with aluminium bromide and with iodine.

The vapour pressure curve OQ is represented by the Clausius–Clapeyron equation:

$$d(\ln p)/dT = \Delta H_v/\mathbf{R}T^2 \tag{1.14}$$

where ΔH_v is the enthalpy of vapourisation, and $\mathbf{R}$ the gas constant. Since ΔH_v depends only weakly on T, this equation can be integrated to give the following approximate equation, which, however, is adequate for present purposes:

$$\ln p = A - \Delta H_v/\mathbf{R}T \tag{1.15}$$

This is the equation of the straight line relating $\ln p$ to $1/T$, with intercept A and slope $\Delta H_v/\mathbf{R}$. Its usefulness here lies in the fact that for any substance for which A and ΔH_v are known, the pressure at any temperature can be calculated. This means one can calculate to what temperature any particular substance may be heated safely in a closed system. If such calculations show that during the operation a pressure above atmospheric can be expected, the securing of all ground-glass joints by strong springs becomes especially important. If calculations indicate that in a planned experiment pressures in excess of ca. 3 atm. are likely to arise, the plan must be altered to avoid this hazard. Because under certain circumstances mixing can be difficult and heat removal therefore may be inefficient, particular attention should be given to the manner in which strongly exothermic reactions are done.

One feature of vacuum operation, which many do not appreciate until they meet it, is the slowness of heat transfer in a good vacuum when gaseous convection is almost zero. This is most often encountered when the inner tube of a cold trap becomes blocked by frozen material. If what is frozen does not react with any of the components of air, the thawing can be accelerated by admitting a small amount of air to both sides of the blocked tube.

'Against the Laws of Physics'. It is not useful to the beginner if the experienced operator pretends that in the operation of a chemist's high vacuum system the observed phenomena can always be explained in simple physical terms, because all too frequently one meets behaviour which is ostensibly in conflict with the laws of physics.

One of the most common phenomena in this category is that gas appears to linger in certain appendages of a vacuum line – or at least the Tesla discharge (see Section 1.3.2.) in these seems to indicate a pressure which is appreciably higher than in the main line. Only if the diameter of the appendage is very small, say 4 mm or less, and the appendage longer than, say, 30 mm, is this likely to be a genuine phenomenon, and then only for a very short time. In other circumstances, it is most likely due to a coating inside the appendage which absorbs certain volatiles or which gives a strong fluorescence under Tesla coil discharge, or both; or there may be a leak in a fused seal which is too small to be detected by the Tesla discharge.

Another common occurrence which 'ought not to be' is the occlusion of 'bubbles of vacuum' in a frozen liquid, and also the persistence of bubbles in a liquid under vacuum contained in a narrow tube, which may be of long duration. When one pronounces on 'what ought and ought not to be' one is almost always thinking of equilibrium conditions, and frequently neglects the effects of surface tension.

1.2.5.2. Distillation.   By far the most common operation which is done on a vacuum line is to move substances, usually but not always liquid at ambient temperature, from one part to another by distillation; that is why a special section is devoted to it. The most common instances involve the transfer of a liquid from one container to another by cooling the receiver below the temperature of the reservoir.

The physical chemistry is basically very simple: If an enclosed space contains a liquid and its vapour, with the liquid at temperature T_1 and a region of the enclosure is at a temperature $T_2 < T_1$, then any liquid film formed from the vapour in that region will have a vapour pressure lower than the liquid at T_1 and therefore there will be a net transport of liquid from the region of higher temperature to that at the lower temperature. This transport will continue until either all the substance has been transferred, or the consumption of enthalpy of vaporisation at the reservoir and its release upon condensation in the initially colder region has equalised the temperatures. The speed and efficiency of this transfer depend on the mean free path of the molecules and for any set of circumstances this is at a maximum if no foreign, uncondensable, molecules such as nitrogen and oxygen, are present. The speed of a vacuum distillation is actually very sensitive to the pressure of any adventitious gas – this means that if such a distillation suddenly slows down, one immediately suspects that the system has sprung a leak, or that a gas which is not condensable at the temperature of the receiver is being formed or released from solution. Such a slowing down may, however, signal something else, especially when one is distilling from a narrow tube: it may then be due to a very cold metastable layer of liquid having been accumulated near the surface by the removal of the enthalpy of vaporisation and poor mixing. The vapour pressure above this is lower than that which corresponds to the temperature of the bulk of warmer liquid below the cold layer and, when mixing sets in, a violent exaporation takes place, usually as a very distinctive, and occasionally destructive, 'bump'.

Further physico-chemical considerations which must never be forgotten are that, as implied above, heat must be steadily removed from the receiving vessel by some form of cooling and heat must be supplied to the source of the distillate. The heating rarely presents a problem, as water baths or hot air blowers are suitable for small vessels, say up to 0.5 l, and heating mantles or tapes for larger vessels.

The cooling needs more careful consideration. From the simple physico-chemical standpoint the temperature of the receiver should be as low as possible, so that the temperature gradient, and therefore also the pressure gradient, should be maximal. However, there are other considerations, mainly safety and cost. Since glassware must never be subjected to unnecessary thermal shocks, one should not subject it to unnecessarily low temperatures. This means that if cooling in ice or a 'mush' at -40 °C is adequate, one should not use solid CO_2 (-78 °C), much less liquid nitrogen (-196 °C). But what is an 'adequately low temperature'? For any substance whose freezing point (strictly, triple point), is below ambient temperature, any temperature below its freezing point is adequate and, because the vapour pressure curve of most solids has a very small slope (see line OR in Fig. 1.1) it is not worth while going below the triple point. Also, the freezing of a liquid may involve very considerable forces and therefore no liquid should be frozen solid in a vessel containing a probe or pocket or electrode.

The freezing of liquids in narrow tubes, such as hanging burettes, is quite an art. If a distillation is relatively slow, the build-up of crystals may form vapour cavities within the cake. Unless the tube is to be filled completely, it is advisable not to undertake an intermediate melting to remove the voids, because these cavities offer a certain safeguard against the vessel being cracked by the expansion of the frozen cake during the thawing.

One of the hazards against which elementary physical chemistry provides no warning is that some liquids, especially alcohols, do not freeze even when condensed far below their freezing point, but form viscous glasses which can be very awkward to handle. The actual choice of cooling medium is dictated in some places by economics (e.g. another department of the same institution may possess a liquid nitrogen generator), in some places by availability (solid CO_2 from a local ice-cream factory may be much easier to procure than liquid nitrogen), and in some by convenience; a good indicator of the skill and experience of an operator is the care which he takes about cooling.

Some liquids, e.g. styrene and nitrobenzene, have such low vapour pressures at ambient temperatures that vacuum distillation from ambient temperature to a cold receiver can be very tedious. The obvious remedy, to heat the liquid will not, by itself, help because the vapour will condense on any and every surface which is below the temperature of the heated reservoir; in other words, the whole vacuum line will be flooded with distillate. A remedy, which is simple in principle but often awkward to put into practice, is to wrap the whole vacuum line with a heating tape adjusted to a temperature *ca.* 10 °C higher than that of the reservoir; for obvious reasons such a set up will be kept as small and simple as possible.

High vacuum distillation is a more elaborate form of what organic chemists call molecular distillation which is usually done over very short paths because the vacua used are generally poor (10−0.1 Torr). Like molecular distillation, its efficiency as a means of separating volatiles from each other (in contrast to separating them from involatile residues) is very low. Under some conditions the distillation of substances of low vapour pressure can be accelerated by continuous pumping. This must be controlled carefully so that only a negligible amount of the compound passes through and out of the condensing vessel. If the pumping is too fast, the distillate will accumulate in the protecting traps and eventually block them.

Some of the more important uses of distillation in high vacuum apparatus which will be encountered later in this book, are: transfer of a liquid by means of a 'cold finger' into a vessel which cannot itself be cooled; blocking access to vessels by means of a temporary, or 'freeze', valve; washing reaction products repeatedly and progressively with a single batch of solvent; washing the interior of ducts free from solids.

The movement of solids in a high vacuum system by sublimation obeys the same physico-chemical principles as distillation, and some useful examples will be found later, e.g. the formation of 'snow-storms' of pure P_2O_5.

1.3. Essential skills and equipment

1.3.1. Skills

1.3.1.1. General comments. The construction of the main vacuum line should, if possible, be done by a professional glass blower, and the actual work with the line will most likely require the construction of some complicated, non-purchasable, glass devices designed by the operator, which will also require the services of a professional, who should therefore be available in-house or nearby.

However, since it is likely that the designer and operator of the high vacuum system will aim at having as few ground joints as possible, many connections will need to be made by fusing on, and disconnections made by sealing off, with a hand torch, and therefore the operator must be familiar with at least the elements of glass blowing. As there are several books from which this skill can be learnt, no general instructions will be given here, but we will present a variety of hints on construction and procedures, which are not generally described in sufficient detail, and the art and craft of finding and repairing leaks will be described in the next chapter.

Without doubt, the most essential characteristics of the scientist–glass blower include persistence, patience and ingenuity. The beginner should never be discouraged by failures, and in moments of anguish and distress he should remember that we are all born knowing nothing, and that what one fool can do, another can do too.

1.3.1.2. The most frequent operation.   One of the earliest operations which the novice glass blower learns is to join together two round tubes. To make a professional-looking, sound joint of this kind it is almost essential to sit at a bench torch and to rotate both parts of the joint-to-be in the flame as shown in glass blowing manuals. There is, however, another way. The beginner should not be ashamed to clamp one part vertically and to hold the other above it, with the orifices just touching, and to fuse the two together by moving the hand torch with a small, hot flame right around the joint-to-be with a smooth, rhythmic circling motion. Holding the upper piece enables the operator to work the joint gently by compression and extension, with appropriate blowing. Another trick, especially useful for tubing of large diameter, or if both pieces to be joined are heavy and/or awkwardly shaped, is to clamp both parts vertically with the ends just touching, and by means of a hand torch and a thin glass rod to fill the gap and make a joint in this way. Of course, when one half of a joint-to-be is part of a fixed apparatus, such a procedure is inevitable (static joint). The making of this type of joint is facilitated by the fact that the difference in surface tension between colder and hotter glass will pull the soft glass towards the cooler region, so that differential warming of the raw joint can be made to even out the thickness of the glass around the jointed region.

Static joints are the subject of a special form of Murphy's Law: Pinholes and cracks appear most frequently in the least accessible places. The reason is that just because the place is accessible only with difficulty, the working of the glass during the making of the joint will have been hampered and therefore probably not well done. That is why the experienced leak hunter will first suspect a leak at the backs of joints made close to any fixtures, i.e. the parts of the joints which are difficult to see, to get the flame to, and to inspect.

To achieve a good static joint it is important to look at it during the making from two directions at right angles; only if both the N–S profile and the E–W profile appear smooth and free of 'chins' is the joint likely to be successful. After completion, every joint should be felt with the finger tips, as these will reveal ridges and pimples which are barely visible: a joint must feel good. The probability of occurrence of a pinhole is reduced very considerably if both the orifices to be joined have been carefully prepared as shown in the manuals.

Whenever glass has been heated to near its working temperature, strains are set up and these need to be released by a slow cooling process called annealing which is dealt with in glass blowing manuals. The larger, and especially the thicker, a glass artefact is, the more important is the annealing; for the thin 'quill' tubing from which phials are blown and which is used to join them to a manifold, it is unimportant, but for bends in and joins to the trunk lines of 8 mm or more diameter it becomes progressively more important. Any pieces of equipment, which have been made or modified by blowing, which are small enough to fit into an annealing oven should be put through an annealing cycle. This usually involves the temperature being raised to ca. 600 °C and brought back to near ambient over 10 h. (An excellent and simple method of removing recalcitrant deposits, especially from intricate pieces, is to put the dirty object through an annealing cycle.) Any pieces which cannot be annealed in an oven (because none is available or the piece is too large, or the worked glass is part of a fixture) should be annealed after the working has been completed by keeping it hot with a brush flame from the hand torch after completion of the glass working such that there is a light pink glow around the piece for a few minutes, and then reducing the temperature slowly over a period of 3–10 min, depending on the size and complexity of the piece. Because the components of a vacuum line are under a pressure difference which may change from 0 to 1 atm and back again at irregular intervals, it is in the interests of safety that as much of the system as possible should be as free from strains as possible. Commercial strain viewers are cheap and show up the location of any severe strains in the glass.

1.3.2. Equipment
The equipment required for the construction, maintenance and use of a high vacuum system comprises:

normal laboratory tools, materials and equipment, including a stable framework, a wide range of borosilicate (Pyrex or other) glass tubing of different diameters and wall-thickness, a range of ground-glass cones and sockets, a set of Dewar vessels of various sizes (these are not only used for the cold traps but also as cold baths for many other operations on the vacuum line) and a hair-drier or heat gun, because the traditional practice of using a 'brush' glame for warming glassware is not recommended except for annealing glass (see previous section) or for flaming out empty apparatus;

the pumps, valves and other appliances discussed in the subsequent sections of this work;

a glass blower's hand torch and a bench torch, both with a range of nozzles, with a supply of gas and oxygen, glass blower's tinted spectacles to protect eyes from the damaging glare of glass when in the gas and oxygen flame, and a good glass-knife, or a tungsten carbide or diamond point for cutting glass;

a Tesla coil for testing the quality of a vacuum and for hunting leaks.

The blowing line should be thin-walled, very flexible rubber or plastic tubing. At the mouth end there should be a mouth-piece of plastic or glass; the author and many of his collaborators, having been pipe smokers, found the bakelite mouth-piece of an old pipe much easier and more agreeable to clasp in the teeth than either glass or soft plastic tubing. Some workers like to insert into the blowing line a glass tube, 3–5 cm long, *ca*. 4 mm i.d. packed loosely with silica gel secured at both ends with cotton-wool pads, to remove spray from lungs or mouth. If one is working on an apparatus where a (mild) explosion may occur, such as the inadvertent 'popping' of a phial included in a dilatometer which is being sealed off (see Chapter 3), it is useful to include a one-way valve in the blowing line as a protection.

The *Tesla Coil* produces a high voltage, high frequency discharge at very low current from a single electrode, usually in the form of a metal wire or a rod which can be moved freely over the apparatus to be tested. This probe should be held some 1–2 cm from the glass in the region where the leak is suspected, usually where a joint has been made. If the pressure inside the apparatus is less than *ca*. 3 Torr, a discharge glow will be produced. Initially, the discharge appears as a single, purple line but as the pressure is reduced, the colour of the discharge changes (see Table 2.1.) and the extent of the discharge increases. At pressures between 10^{-2} and 10^{-3} Torr large sections of the apparatus will become illuminated by the ionised gas; such discharges can be seen better in a dim light, but there should always be enough light in the laboratory to see what one is doing. Below about 10^{-4} Torr the discharge within the apparatus disappears; hence the term a 'black vacuum'. This colour guide to the pressure assumes that the original gas was air, since the colour depends on the nature of the gas remaining, and most solvents have their own typical discharges which can mask the colour due to any remaining air. Between 10^{-3} and 10^{-5} Torr the glass walls start to fluoresce with a range of colours depending on the nature of the chemicals adsorbed on the surface.

If there is a pinhole in the glass which is sufficiently large, it will attract the discharge from the Telsa coil in the form of a long blue spark which ends on the glass surface at the pinhole in a sharp, blue-white point; fortunately, leaks too small to be detected by the Tesla discharge are extremely rare.

Two warnings   (i) The Tesla coil is a high voltage generator and as such is a transmitter of weak radio signals. These can disturb sensitive electrical equipment and are usually picked up by chart recorders in the vicinity. (ii) The discharges can also destroy certain semi-conductor devices such as operational amplifiers; therefore any equipment containing such elements must always have electrical shielding if it is to be used near a vacuum line.

There are at least three sets of circumstances where the Tesla coil discharge cannot be used to locate leaks:

(1) Since the spark goes to metal in preference to glass, the discharge will not locate a leak at any glass–metal junction, such as a glass–metal joint or where a wire is sealed through glass, or where the glass is near metal, such as at a clamp.

(2) The spark punches a hole through thin glass ($< ca.$ 0.5 mm) and can therefore not be used on breakable phials or break-seals.

(3) The discharge must not be used to test for a leak from the outside along the plug of a PTFE tap or from one side of the tap to the other, because if there is a leak, the spark burns a track into the surface which ruins it irreparably.

The methods of locating leaks in these circumstances are described in Section 2.3.2.

1.3.3. Some unconventional devices

Some items not usually described in elementary glass blowing texts, but frequently useful in vacuum apparatus, are *magnetic seal-breakers, universal joints*, and *metal wire seals*. Therefore their construction is described here.

1.3.3.1. Magnetic seal-breakers. For many purposes, such as moving or breaking phials of reagents inside a vacuum system or opening break-seals, it is useful to have an iron rod inside a vacuum apparatus which can be moved by a magnet. PTFE-covered magnets, usually employed as stirring bars, are suitable for many such operations, but they are easily scratched and are then difficult to clean. French workers have usually used stainless steel ball-bearings, but the impact of steel on glass is unnecessarily harsh and risky. A convenient alternative is an iron bar enclosed in glass, which is known as a *glass-coated seal-breaker* or magnetic hammer.

The sequence of operations for making such a seal-breaker is shown in Fig. 1.2. First, select a glass tube of wall-thickness *ca.* 1–2 mm which fits easily (not tightly) over an iron rod; instead of a single iron rod, a bunch of smaller rods (decapitated nails) can also be used. It should have an outer diameter (o.d.) *ca.* 2 mm less than the i.d. of the tube in which it is to travel. A constriction is prepared in the glass tube (Fig. 1.2.(a)) and a small piece of high temperature ceramic wool (asbestos substitute) is pressed into the tube at this constriction. Before use the wool must be cleaned by soaking it in hexane, thoroughly dried and heated to redness briefly in a flame. The iron rod is then dropped into the tube followed by another small plug of ceramic wool. After pressing the contents together with a glass rod, the end of the tube is sealed to a round end, care being taken not to heat the contents of the tube more than absolutely necessary (Fig. 1.2(b)). Finally, the tube is attached to a water pump, evacuated and sealed at the constriction. To effect

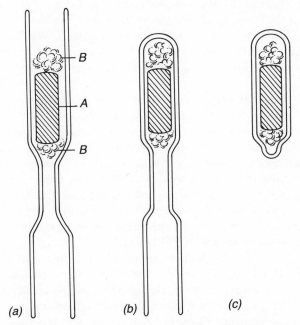

Fig. 1.2. The construction of a magnetic seal-breaker. *A* iron core or bundle of nails, *B* ceramic wool (asbestos substitute). For details see text.

the seal one first warms the part containing the iron core and then moves the heat to the constriction and seals this as rapidly and neatly as possible (Fig. 1.2(*c*)); the breaker should not be heated after it has been sealed. The seals at both ends of the breaker must be made with all possible care, because of course pinholes can occur there and once the breaker is in place inside the vacuum system, they are almost impossible to detect, but can ruin the proper operation of the system.

1.3.3.2. The universal joint. If two parts of a vacuum system are to be joined whose distance apart and relative positions are not fixed exactly and which must be separated and joined again repeatedly, such as a reaction vessel which needs to be detached for clearing and cleaning, three kinds of junction are available. One is a thick rubber or plastic tube, which is generally unsatisfactory. The second is a corrugated metal tube, also known as bellows tubing, both ends of which carry metal cones machined accurately to fit either B14 or B19 glass sockets.

The third device consists of four glass right-angled bends whose arms are *ca.* 3–5 cm long and which carry a cone at one end and a socket at the other. The author has used such links both with B10 and B14 joints. When the four links are assembled together they provide a highly flexible, adaptable junction which, in contrast to the bellows tubing, does not impose a mechanical stress on the parts which are joined. It is often useful to fasten successive links to each other by small steel springs engaging hooks fixed to the ends of the links near the joints. Similar devices can be made, with obvious improvement in flexibility, from ball-and-socket joints.

1.3.3.3. Metal wire seals. If very aggressive-corrosive materials must be handled in a high vacuum system comprising cone-and-socket or flanged joints and none of the three types of grease (hydrocarbon, silicone, fluorocarbon) are suitable, one can resort to metal wire joints. These consist of a loop of wire laid around the cone, or on the flange, with the free ends laid simply one across the other; a butt-weld of the two ends is definitely not necessary. The cone with its wire loop is rammed into the socket with a very firm action and a very slight twist. This will give a vacuum-tight seal with a normally ground cone and socket, but for the easiest and most reliable joint a polished cone and polished socket should be used.

The most suitable wires for this purpose are pure lead, indium and gold. The Pb and In wires should be between 0.1 and 0.4 mm thick, the Au *ca.* 0.05 to 0.1 mm. Under presure, these metals are deformed plastically and adapt themselves to the contour of the joint. There will momentarily be the greatest pressure at the point at which the ends cross over and thus the greatest deformation will occur there until a uniform thickness is attained. If such a joint resists dismantling, it can usually be freed by heating it gently with a brush flame or hot air gun.

1.4. General comments on safety

Twenty years ago there would have been little need to have a section on safety in a book designed to help the practising chemist fulfil his experimental plans. It is not chemists becoming more careless or their science becoming more dangerous that requires such a section in a book written at this time. The reason is that whereas in the past research chemists, and particularly those in academic institutions, were almost forgotten in terms of legislation, this is not so today. Recent legislation in most industrialised countries has made the research chemist legally responsible not only for his or her own safety, but also for that of the other people who come into the immediate environment of the experiments being done. This legal obligation implies that the research chemist must take more care and be more aware of the possible dangers associated with his experimental programme. Throughout this book the reader will find mention of particular points related to the safe usage of individual techniques, procedures and pieces of apparatus, but several general points related to the operation of equipment under vacuum are worth noting in addition to the general rules related to safety in the chemical laboratory.

Although there is no danger of condensing oxygen within equipment which is maintained well below atmospheric pressure, any equipment that has been isolated from the pumping system should be checked very carefully for leaks *before* cooling it with liquid nitrogen. If liquid oxygen has condensed in any apparatus which also contains organic material there is a serious explosion hazard. Such apparatus should be kept at the temperature of liquid nitrogen, isolated from the vacuum line and removed to a safe place where it can be opened to the atmosphere and allowed to warm up *slowly* to ambient temperatures.

In addition to the explosion hazard presented by condensed liquid oxygen, it must always be remembered that the vacuum line in operation is a closed system and that liquids condensed in a reservoir under vacuum will attain their own vapour pressure at any temperature which they subsequently reach. Glass is a relatively strong material under tension and even stronger under compression, but it is very brittle. Flasks containing liquids of low boiling point, such as sulphur dioxide, 2-methyl propene or boron trichloride, are under several atmospheres of pressure at room temperature; the slightest knock can cause a flaw in the glass, and if it is under stress an explosion may ensue. Such flasks should be strapped with sticky tape and surrounded by a cage of wire gauze. The purpose of the latter is to allow the shock waves to escape but to contain the glass fragments in the event of an accident. It also serves to protect the flask from accidental knocks and the explosion of a neighbouring piece of equipment. For this reason all flasks larger than *ca.* 0.5 l should be enclosed in wire mesh.

Vessels under vacuum do not explode but implode, and the consequences may be significantly worse than the pressure difference of 1 atm suggests. Implosions are, however, very rare and usually result from apparatus being incorrectly designed or handled. The author, with 40 years of laboratory experience, has never seen an implosion.

Because flat glass surfaces are weaker than outwardly convex ones, and outwardly concave ones are even worse, no such surface of more than a few square centimetres should ever be evacuated on the convex side, and the wall-thickness must, of course, be adequate.

Ring-seals of all kinds tend to be mechanically weak and sensitive to thermal shock. Tungsten-through-glass seals and pockets for thermocouples or conductivity probes are the most frequent instances of ring-seals on vacuum equipment, and these should never be subjected to extreme cold.

A warning about glass blowing: The flame of a glass blowing torch should never be directed into the neck of a flask because the mixture of air and unburnt gas is very likely to explode violently.

Since one of the major reasons for working with a vacuum line is the sensitivity of the materials being handled to oxygen or water, it follows that many of these materials present severe hazards should an accident occur. Of course, these chemicals must also be introduced somehow into the vacuum system, and this operation must be carefully thought out with special regard to the safety precautions. The chemist working with hazardous compounds must think about what to do if the vacuum system fails. Rapid removal of the affected part of the apparatus to a safe place is often the best answer, but the individual solutions to such problems must be worked out before starting operations. Do not wait for an accident to happen before considering how to dispose safely of the contents of the vacuum system.

Caveat lector: Think before you start!

Finally, it is obvious that the chemist working with h.v.t. must wear protective spectacles at all times, and rooms containing vacuum equipment must be appropriately labelled.

There is a vast literature on safety in laboratories, and in this the publications of the Royal Society of Chemistry are prominent. We give below details of works recently issued or reissued:

(1) *Hazards in the Chemical Laboratory* 4th Edition. Edited by L. Bretherick.

(2) *Guide to Safe Practices in Chemical Laboratories* (no author).

Details are obtainable from: The Royal Society of Chemistry, Sales and Promotion Department, Burlington House, London. W1V 0BN, England. Telephone (01) 734 9864. Telex: 268001.

Reference

S. Dushman, *Scientific Foundations of Vacuum Technique*, 2nd Edition, Wiley, New York, 1962.

2   The main vacuum line

2.1. In the beginning

In this chapter the reader is provided with the information necessary to decide what form his vacuum line should take, and how to construct it, test it, use it, and eventually dismantle it.

2.1.1. Guiding principles

Before discussing the individual units which together constitute the main vacuum line, there are several considerations concerning planning which, although they may appear to be obvious, are worth emphasising here.

The proposed experimental plan must be thought out carefully, so that the maximum amount of information can be obtained with the minimum of effort. To this end the intelligent chemist will investigate whether h.v.t. can provide any short cuts. For example, a compound being studied may be stable enough under vacuum to be stored for long periods, whereas it would deteriorate rapidly if prepared on the bench and kept under nitrogen. In that case, the synthesis of a larger batch of material under vacuum may be more profitable than several smaller-scale syntheses on the bench. Another example is the possibility of making more than one analysis on each reaction mixture. For instance, if NMR measurements are planned, it is always worth considering whether additional information would be gained from measuring the electrical conductivity or perhaps the UV spectrum of the same solution. The appropriate modifications to the vacuum apparatus would certainly be less time-consuming than having to make up new solutions for the additional analyses. The modifications would include provisions for diluting the solution which is suitable for NMR measurements to the low concentrations required for e.g. UV measurements. This is an appropriate occasion for emphasising that if one is dealing with a reasonably volatile solvent, it is just as easy on a vacuum line to concentrate a solution as it is to dilute it.

Another point worth considering in detail before starting to build a vacuum system is the nature and quantities of solvents which will be required. Careful purification of solvents is a prerequisite for most h.v.t. and it is usually wasteful and tedious. Therefore, if large amounts of particular solvents will be required, then it is better to purify in larger batches and to incorporate correspondingly large solvent reservoirs in the system.

The secret of successful h.v.t. is careful planning, and the choice of the appropriate apparatus and technique will become easier with practice. The best-planned vacuum line will be usable for operations for which it was not designed originally, which means that it is versatile and adaptable. It should be suitable for preparative, kinetic and analytical work, and it should not be so complicated that cleaning requires extensive dismantling and reconstruction. The next three chapters will provide the reader with the information necessary to design the right system for his particular purposes.

The first stage in the planning of a new line is to list the proposed operations which the experimental work will entail. Next, a drawing should be made, showing which piece is connected to what (the equivalent of a circuit diagram in electrical work), and concurrently a list of the required components is drawn up with rough specifications e.g. 'three glass taps, 9–12 mm bore'. It is useful to combine idealism with realism and to put together two lists – one of the most desirable items and one of those items which are the minimal requirements – and to cost both, and to mark what is available and what must be procured. At this stage, pricing by means of suppliers' catalogues is, of course, essential. The description and discussion of the various types of each category of component which follows this section are intended mainly for the beginner, but no doubt several old hands will find some new, useful information; the author knows only too well that long practice can make one set in one's ways and resistant to innovation ('better the Devil you know...'). This is an excellent reason for bringing new blood into old laboratories.

A diagram of a basic vacuum line, with all its essential parts, is shown in Fig. 2.1. Its constituents and its mode of operation will be explained in the following sections of this chapter.

2.1.2. The foundations: the bench and the frame

What the foundations are to a house, the bench and frame are to a vacuum line, and as with the siting of any building, the location of a vacuum system is important.

The proposed site of the vacuum line should not be along a major laboratory thoroughfare, and it is essential for the operator to be able to move freely in front of the line, and access to the rear of the line, although not essential, can be extremely useful. The vacuum line must not have a window behind it, because if it does, glass blowing work on it becomes very difficult, and the Tesla coil discharge will not be visible, at least during the hours of daylight; the best background is a matt black wall or screen.

The ideal support for a vacuum line is a heat-resistant, chemically inert table some 40–60 cm from the floor, 40–60 cm wide and at least 1.3 m long. The height should be such that a rotary pump can be fitted underneath, the legs should be screwed to the floor, and the space below the table should be open to access from all sides. Bolted firmly to the table and to at least one wall should be a frame of angle-iron carrying a grid of iron rods which will be the main structure supporting the glass line. This metal framework should be earthed, preferably by an electrician. The bolting to the wall must be such as to prevent any vibrations from pump or stirrer motors. The working surface should have a gallery *ca.* 1.5 cm high to contain any spilled liquids, especially mercury.

In designing the layout of a new vacuum system it is important to take the height of the operator and the frequency of probable use of the individual parts into consideration: A well-designed line will not require the operator to spend long periods either kneeling on the floor or standing on steps. It is also important to take into consideration the position of the essential services (gas, oxygen, water and electricity) in the laboratory; it is undesirable that the operator should have to reach frequently though the apparatus to plug in a piece of electrical equipment or to turn the water on or off. Such actions are time-consuming and constitute unnecessary hazards.

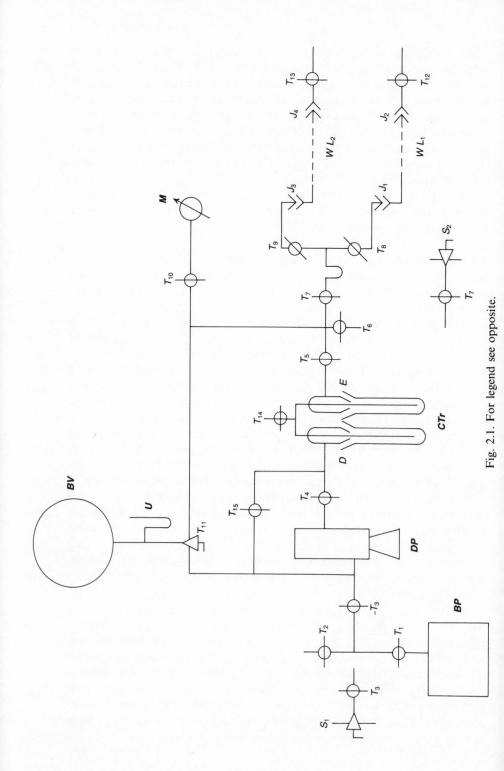

Fig. 2.1. For legend see opposite.

If at all possible, a laboratory bench or a table *ca.* 1 m high should be available at right angles to the vacuum frame, for ancillary equipment such as oscilloscopes, electrical instruments, microcomputers etc. and there should be within easy reach of the operator a bench or table for Dewar vessels, reagents, etc.

The ideal arrangement is a very individual matter, and each operator should devote time to establishing comfortable working conditions which suit his shape, size, personality and chirality.

2.2. The individual components

2.2.1. Pumps
The pumping system is the most important part of the vacuum line, it is certainly the most expensive, and it should therefore be chosen with care. Because the pumping system on a chemist's vacuum line must be robust and capable of removing large quantities of gases, often in repeated cycles, sorption pumps, getter ion pumps, and sublimation pumps are generally unsuitable and are therefore not discussed in this book.

2.2.1.1. The rotary pump   The principle of operation of a typical rotary pump is indicated in Fig. 2.2. The cylinder A rotates within the cylinder B with an eccentric motion. The spring-loaded vane C maintains a good seal separating the inlet D and the outlet E. A pumping action is achieved because as the cylinder A rotates, the space F is enlarged and the residual gas in the system diffuses through the inlet. At the same time the space on the outlet side of the vane C is reduced and the gas in it is forced through the outlet E via a one-way valve which prevents any back flow of the lubricating oil when the pumping is stopped. Most modern pumps are also fitted with a safety valve on the system side to prevent oil being sucked into the vacuum line if the pump is switched off (or the electricity supply fails) while the system is under vacuum. Most commercially available rotary pumps contain two units

Fig. 2.1. The vacuum line. The parts of this diagram are not drawn to the same scale. See also Fig. 2.14.

BP backing pump, *DP* diffusion pump, *CTr* cold traps, *BV* ballast volume of 5–10 l capacity, *U* indicator manometer, *M* accurate manometer.

WL_1, WL_2 alternative working lines. The second one is, of course, optional. J_1–J_4 cone-and-socket joints, at least B.19, preferably B.24. J_1 and J_2 can, of course, be replaced by fused junctions. T_1, T_3–T_{13} hollow key vacuum taps of not less than 8 mm key-hole diameter. T_2 and T_{14} can be small solid-key taps. The taps T_2, T_3 and T_8, T_9 can be replaced by the 'either–or' two-way taps S_1 and S_2.

A by-pass connecting the high pressure side of the *DP* (between T_3 and T_{11}) with its low pressure side (between T_4 and D) and comprising a tap T_{15}, can be useful for protecting the *DP* if the line must be 'let down to air' frequently whilst the *DP* is hot, and is essential for 'hard' evacuation of *BV*.

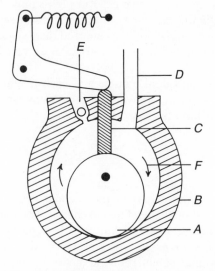

Fig. 2.2. Diagram of the rotary oil pump. Its functioning is explained in the text.

of the type described here, totally immersed in oil within an outer casing. These units usually operate in series and are driven by a single shaft.

The exhaust from a rotary pump, especially if it is being run in the 'ballast' mode, i.e. pumping a fair quantity of air or other gas, is an aerosol of oil in the gases from the line. A variety of filters is now available commercially for cleaning exhaust gas, but a good additional safety precaution is a wide tube fitted to the outlet so that the gas stream can be vented to the nearest fume-cupboard or window. The pumping efficiency of a rotary pump drops off rapidly below ca. 10^{-2} Torr even under optimum conditions, and such pumps are therefore usually installed as a back-up to a more efficient high vacuum pump, such as a diffusion or a turbomolecular pump.

2.2.1.2. The Cole pump   A new type of pump, invented by M. Cole (Cole, 1987a) has recently been introduced for handling corrosive and condensable gases. The principle of operation is this: A spinning drum is partly filled with an inert liquid of low vapour pressure and a stationary probe is held in the circulating liquid. The liquid moving over the top surface of the probe separates at an abrupt change of profile to form a vacuum cavity. A hole or slot under this cavity is connected via the probe and the stationary central hub to the vessel to be evacuated. In operation, gas is drawn into the vacuum cavity at the top of the foil and carried downstream in the rotating liquid in the form of gas bubbles. The centrifugal field in the liquid causes the gas bubbles to migrate towards the empty space in the centre of the drum, where they burst. The gas thus released is vented to the atmosphere through suitable holes in the central hub. Commercial versions of these pumps are at present limited to an ultimate vacuum of about 10^{-3} Torr although 10^{-4} Torr has been achieved in laboratory prototypes (Cole, 1987b).

Because there are no rubbing components in this type of pump, it is easy to make them of corrosion-proof materials, and because there are no tight clearances it is also easy to run them at 120 °C. This allows operation above the boiling point of water and most laboratory solvents. Operation at this temperature together with a 'ventilation' facility (comparable to gas ballast in rotary vane pumps) allows the Cole pump to handle vapours at input pressures of 1 atm compared with about 20 Torr for gas ballasted rotary vane pumps.

Backing a rotary vane pump with a Cole pump (evacuating the oil box of the rotary vane pump) provides a combination which can achieve the high ultimate vacuum of a rotary vane pump combined with the vapour handling capability of a Cole pump.

2.2.1.3. Diffusion pumps The principle of a diffusion pump is best demonstrated by examining the way in which a simple mercury diffusion pump, such as that shown in Fig. 2.3, works. The pressure is first reduced to approximately 5×10^{-2} Torr by a backing pump, since above this pressure diffusion pumps are very slow. The mercury is then heated so that it boils vigorously and a stream of mercury moves rapidly from the boiler B to the walls of the condenser C. The gas molecules in the inlet area of the pump become entrained in the heavier mercury vapour and are accelerated towards the backing pump connected to E. Provided that the backing pressure is maintained below a critical value such pumps can achieve very high pumping speeds and low ultimate pressures ($ca.$ 10^{-6} Torr).

Mercury diffusion pumps are normally constructed from quartz or heat-resistant glass and are therefore a possible source of hazard should they break, especially whilst they are hot. However, during over 40 years of working with such pumps, the author has neither experienced nor heard of such an accident. The major real disadvantage of mercury pumps is the relatively high vapour pressure of mercury at room temperature ($ca.$ 10^{-3} Torr), which makes its necesssary to ensure that the cold traps prevent efficiently the mercury vapour from diffusing forward into the line.

More modern diffusion pumps are operated with a range of synthetic oils having room-temperature vapour pressures of between 10^{-5} and 10^{-9} Torr. Although many of the oils previously used in oil diffusion pumps were not particularly stable to chemical attack at the normal working temperature of the diffusion pump, the oils available today (based on a variety of materials such as naphthalene, poly(phenyl ether), or silicones) are generally stable to oxidation at their normal working temperatures and many are particularly suitable when contact with more aggressive materials cannot be ruled out.

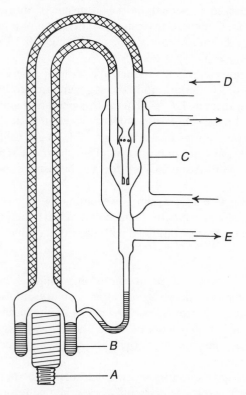

Fig. 2.3. Two-stage glass mercury vapour pump, the 'work horse' of hundreds of laboratories for many decades. *A* electric heater, *B* boiler for Hg, *C* water condenser, *D* to vacuum line, *E* to backing pump.

Another advantage of the oil diffusion pumps is that, by virtue of their multi-stage design and the greater molecular weight of the oil compared to mercury, they can achieve greater pumping speeds at higher backing-pressures and they can also achieve lower ultimate vacua; in other words, they are more efficient. They do, however, have two disadvantages compared with mercury diffusion pumps: (1) Contamination of the pump fluid is a much more frequent problem than with mercury pumps, which means that the pump fluid must be changed more often; and (2) the forward diffusion of oil into the system must be prevented carefully whereas traces of diffused mercury can often be tolerated in the main line. To reduce the forward diffusion of oil, most oil diffusion pumps are fitted with a series of baffles, but it should not be assume that such baffles eliminate the problem completely. Another way of reducing the diffusion of oil vapour into the vacuum system is to pack one of the cold traps loosely with glass wool; this increases the cold area, decreases the main free path and thus enhances the efficiency of condensation.

All diffusion pumps must be fitted with an electrical cut-out system to prevent overheating of the oil or mercury should the cooling system fail. No diffusion pump must ever be opened to the air whilst the contents are hot, because this would cause oxidation of the oil or mercury; silicone oils are less sensitive in this respect but they are not totally inert.

2.2.1.4. Turbomolecular pumps   The principle of a turbomolecular pump is very simple: A particle that collides with a much larger moving body comes away from the collision with an additional momentum in the direction in which the larger body was travelling. The total momentum of the particle is then made up of a component due to its thermal energy and that due to the collision. If the larger body forms the wall of a space in which the particle is travelling and the collisions take place at a pressure low enough for molecular flow to predominate, then collisions with the wall will occur much more frequently than collisions between particles. The net effect of such a molecular pump is that the particle is accelerated in the direction of movement of the wall. Modern turbomolecular pumps are made up of a series of turbine-like rotors and stators orientated so that the stators form a mirror image of the rotors. Each individual blade of the rotors operates as a single molecular pump and, although the individual rotors are rather inefficient due to back-diffusion between the rotor and the case, modern pumps, with 20 or more rotors, can achieve ultimate vacua down to 10^{-10} Torr. The turbomolecular pump requires a backing-pump in order to bring the pressure in the system to *ca*. 10^{-2} Torr where the turbomolecular pump starts to operate efficiently.

The major disadvantage of a turbomolecular pump is its low efficiency with respect to hydrogen gas. This low efficiency is inherent in the nature of the pump; the pumping speed is proportional to an exponential function of the square root of the molecular weight of the gas being pumped. For most applications, however, this lack of efficiency is of little importance since, even if the total pressure in the system were due to hydrogen, when this falls to less than 10^{-6} Torr, the effect of hydrogen on the chemical system would have to be catastrophic for it to be noticed. In terms of the speed of operation in the region 10^{-3}–10^{-4} Torr there is little difference between a turbomolecular pump and a diffusion pump when both are in operation with similar backing vacua, but the pumping speed of a diffusion pump is less sensitive to the efficiency of the backing pump. On the other hand, the turbomolecular pump requires no warm-up time and can be operating at full efficiency as soon as the backing pump can bring the pressure in the region of the pump to below *ca*. 10^{-1} Torr.

In contrast to the diffusion pump, the turbomolecular pump can be connected in series between the backing pump and the system. Commercially available pumps usually have a cut-out mechanism which turns the pump off when the resistance in the system, due to increased pressure, slows the rotation of the pump to about 90% of its top speed. This is a safety precaution, since at high speeds and high pressures the cooling system cannot cope with the heat developed and the pump then overheats rapidly. Those parts of a turbomolecular pump which come into contact with the gases being pumped are generally made of high quality stainless steel, and provided that adequate precautions are taken to ensure that no chemicals which corrode stainless steel, or small particles can get into the pump, such pumps are extremely robust, easy to operate and require very little maintenance.

Until recently the major disadvantages of turbomolecular pumps were their size, the noise which they generated and their price. However, the most modern types are small enough to hold in one hand and make no more noise than a rotary pump, and they are not significantly more expensive than diffusion pumps. A very important advantage of the turbomolecular pump is that there is no oil to diffuse up-stream into the system.

2.2.1.5. The choice of pumps   The choice of which vacuum pumps to use, or to buy if they are not available in stock, is more difficult. Old pumps can still be very efficient, if they have been treated well and maintained properly. The effect of increasing the nominal speed of the pump on the rate of evacuation of the system can be derived from the more detailed discussion in Chapter 1, but a full analysis of the desirable pump size for a particular system is generally not necessary for a chemist's vacuum line. The maximum pumping speed is usually controlled by the conductance of the system and the short term, ultimate vacuum is limited by the rate of degassing of the glass surfaces. Even with the largest and most expensive pumps, a vacuum much better than 10^{-4} Torr (measured just before the high vacuum pump) cannot be expected without several hours of evacuation. When buying new pumps it is advisable to check the diagrams showing pumping speed vs. pressure rather than to rely on the nominal pumping speeds quoted, and to ensure that the high vacuum pump will not suffer a reduction in efficiency because of an inadequate backing pump. This problem is best avoided by buying both pumps from the same manufacturer, which also ensures that fittings will be compatible. For the same reason, it is worth checking what pumps are in use by other groups in the same institute so that a large supply of spare parts need not be acquired by an individual worker. With these considerations in mind it is generally best to buy the most powerful and robust pumping system which one can afford.

2.2.2. The cold traps
The purpose of the cold traps (see Fig. 2.1.) is two-fold: (1) to trap volatile materials from the line on their way to the pumps and thus to protect the pumping system: and (2) to trap the vapour and any pumping fluid before it can enter the line by back-diffusion.

The most common coolant used for the traps is liquid nitrogen (b.p. $-195.8\ °C$). In comparison to liquid air, which was previously used as coolant, liquid nitrogen has the advantage that it does not support combustion, whereas liquid air can start fires, or even explosions if in contact with organic materials. This property of liquid air derives from its oxygen content (b.p. -183.0 °C) and it is because oxygen can be condensed at the temperature of liquid nitrogen that the traps should *never* be cooled with liquid nitrogen when the system is not under vacuum or at least being pumped out. If for some reason liquid oxygen has condensed in the cold traps, the first sign is that the vacuum obtained is very poor, even with the main system closed off: it appears as if there is a leak somewhere on the pump side of the main line. If this happens, the Dewar vessel containing the liquid nitrogen should be lowered carefully whilst the pumps are still running and the traps inspected. A larger quantity (*ca.* 0.5 cm³) of condensed oxygen will appear as a pale blue liquid. If the traps do not contain any organic material and liquid oxygen is suspected, then the traps should be allowed to warm up slowly with continuous pumping. If the cause of the problem is condensed oxygen then the pressure at the pumps will rise as the temperature of the traps rises and should fall to an acceptable level after a few minutes, when the oxygen has been pumped away. Should liquid oxygen be suspected when the traps also contain organic material, the best approach is to close the traps off from the system, vent them to the air and remove them, still cooled in liquid nitrogen, to a place where they can warm up slowly without creating a safety hazard.

For some purposes solid carbon dioxide ('dry-ice'), sublimation temperature -78.5 °C or mixtures of dry-ice and acetone (temperature -78 to -95 °C) are used as coolants. These are obviously not as efficient as liquid nitrogen and they should not be used with chemicals which have an appreciable vapour pressure at the appropriate temperatures.

The two main designs of trap are shown in Fig. 2.4. The type shown in Fig. 2.4.(*a*) is the most common and has the advantage that it can be cleaned easily. The arrangement shown is that which leads to the minimum danger of the condensing materials blocking the traps. Should the traps become blocked, the coolant should be removed, the traps isolated from the system (Taps T_4 and T_5 of Fig. 2.1), and vented via tap T_{14}, and the blocked portion carefully warmed with a hair drier until the traps can be dismantled (joints D and E of Fig. 2.1) and cleaned. Special precautions must be taken when the condensed material might be sensitive to air or water, and also when the material blocking the traps is ice. A useful precaution when air- or water-sensitive materials may condense in the traps is to use the venting tap T_2 which enables one to vent the pumps without drawing air through the traps. The contents of the traps can be protected under a blanket of dry nitrogen whilst the traps are dismantled and their contents disposed of, by letting down the vacuum to a supply of dry nitrogen through T_2, T_6 or T_{14}. Another arrangement of traps, useful when the materials which may condense are extremely reactive with moist air, includes a third trap. The trap closest to the

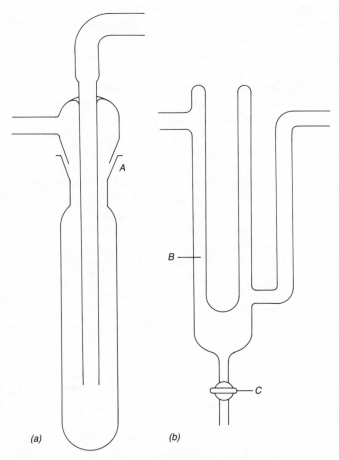

Fig. 2.4. Two types of cold traps. *A* is a B.24 or B.29 joint, *B* is the well for liquid nitrogen or other coolant, *C* is a 3–5 mm bore solid-key tap.

vacuum line is filled to within 2 cm of the opening of the central tube with an appropriate diluent so that as the traps are allowed to warm up at the end of the experiment, the frozen reactive material drops slowly into the diluent; e.g. with aluminium alkyls, the first trap can contain toluene in which aluminium alkyls react slowly with the water present and are rendered harmless. Where larger quantities of material are likely to be condensed, the trap design shown in Fig. 2.4(*b*) is particularly useful. Such a trap can be isolated from the system, warmed quickly with a hair-drier, and the contents removed via the tap. Such traps are difficult to clean but have the advantage that no Dewar vessel is required, the coolant being introduced into the trap in the same way as for a cold finger. For greater ease of dismantling and cleaning, the arms of trap (*b*) can end in vertical cones which engage corresponding sockets on the vacuum line, preferably at least B.29.

2.2.3. The main manifold or trunk line
The trunk line is the pumping route to the various lateral and ancillary lines and should have as large a diameter as possible; the largest diameter tubing that can be blown conveniently by a practised amateur glass-blower is *ca.* 15 mm i.d. (see section 2.3.1).

Several tips are worth noting with respect to the trunk line:

(1) In the schematic diagram of the vacuum line (Fig. 2.1) a U-bend is shown between the main tap T_7 and the manifold, which has two functions: (*a*) It lends an extra element of mechanical flexibility to the system by absorbing small movements at the traps or along the main manifold (WL_1 or WL_2) which might otherwise lead to fracture; and (*b*) it acts as a sink for non-volatile residues in the line and for grease which may be washed away from the taps T_7, T_8 and T_9.

(2) The outlets from the working lines should always sprout from their upper side because outlets from the underside act as sumps for non-volatile material travelling along the line, such as substances in solution if the line is flooded or powders carried into the line by a gas stream during the early stage of evacuating a powder in a reservoir; both types of accident are not uncommon. If an outlet from the underside of the manifold ends in a tap, as they almost always do, then if this is of glass any organic liquid collecting over it will seep into the lubricant with unfortunate consequences, and if the tap has a PTFE key, the liquid collected above it, cooling whilst being pumped off, will cause the PTFE key to contract and leak.

(3) The main manifold should be as long as possible to provide the greatest possible adaptability, and its end should be closed by a B.24 or B.29 socket and cone fitted with a tap; this makes it possible to clean the line easily, and the outlet from the tap can have the glass blower's tube attached to it. Cleaning the line is also made much easier if it slopes downwards a few degrees towards the free end. It is most convenient to operate if mounted at about chest height for the operator.

2.2.4. Taps and valves
In this section only those taps and valves will be considered which are suitable for repeated use. Break-seals and seal-off points which can be considered as once-only valves are discussed in Section 3.2. Before discussing the different types of taps and valves there is a safety rule which applies to all taps which are operated by turning a key: *Always use two hands*: one hand to hold the tap barrel and the other to turn the key.

2.2.4.1 Glass taps   Glass taps should only be used at the pumping end of a vacuum system or where there is no likelihood of them coming into contact with liquids since, whatever grease is used, it will tend to be washed away from the tap, and the grease absorbs organic and other vapours which must then be pumped off again. Except for air vents, only hollow glass taps should be used in vacuum systems. In view of the drastic effect of constrictions on the pumping speed (see Section 1.4.) taps with the largest available bore should always be used, despite their cost. Modern glass taps usually have interchangeable keys, but the fit of a replacement key is frequently not perfect. This difficulty can be remedied easily: A fine carborundum powder is mixed with glycerol to a paste with the consistency of mustard and applied sparingly but evenly to the new key. The key is then gently lowered into the barrel and rotated several times with little or no pressure. The paste is then cleaned off and the fit checked. If the fit is still not perfect, the procedure can be repeated with the application of a slight pressure during the rotation of the key.

Glass taps require greasing *before* any vacuum is applied. The answer to the question of what grease to use depends on the nature of the chemicals being used in the system. Three types of grease are in common usage: Those based on hydrocarbons such as Apiezon, those based on silicones, and those based on perfluorinated hydrocarbons. It is worth checking the manufacturer's description of any grease and considering the nature of the chemicals to be handled in the system before applying any grease. In this respect, it is especially worth noting that most reactive metal halides such as BF_3 and $TiCl_4$ react with silicone greases to yield an excellent cement; therefore such greases must never be used where there is any chance of contact with a metal halide of this type. Most of the hydrocarbon greases become more viscous with age because they loose their lower molecular weight components. They are also degraded rapidly by contact with anhydrous strong acids and by reactive metal halides. Lastly, all greases tend to contain some dissolved gases and water and therefore a freshly greased tap in an otherwise aged system will at first give a poorer vacuum. This difficulty can be avoided by using grease from a stock which has been evacuated for several hours whilst being heated in a boiling water bath; this treatment is almost essential for silicone greases which usually contain much water. An important consideration is the temperature coefficient of viscosity. If large temperature fluctuations are expected, e.g. because of 'economical' space heating in winter, silicone greases are preferable to hydrocarbon and fluorocarbon lubricants.

The most effective way to grease a tap is shown in Fig. 2.5. Having applied a small amount of grease as shown, the key is lowered slowly into the barrel in the 'open' position and squeezed down gently. Only after the key is sitting firmly in the barrel is it rotated through one or two complete turns. Finally, a vacuum is applied to the tap in the 'open' position and it is again turned. The optimum amount of grease is that which provides a transparent film without streaks over the whole area of contact between key and barrel, without any excess being squeezed out above or below the key, or into its holes.

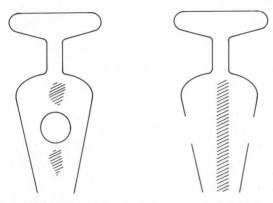

Fig. 2.5. How to grease a tap key. As shown, the grease (shaded areas) is applied in two small patches above and below each hole and in two streaks mid-way between the holes.

The larger the tap, the greater the surface area of contact between key and barrel and therefore the greater is the friction, so that turning large taps requires more force. If the tap is very stiff it should be warmed gently and evenly with a hair-dryer before turning. Never turn a stiff tap in one continuous motion because the pressure applied will be increased involuntarily and uncontrolledly during this operation and this can easily lead to wrenching the whole tap from its connections. Therefore it is better to turn through a small angle and then release the grip and start again. If a glass tap is really stiff, it is better to vent the system and regrease it. A tap that is completely stuck can often be freed, after it has been vented to atmospheric pressure, by warming it evenly with a 'brush' flame or hair drier and tapping it gently with a cork ring or a piece of wood. A stuck key that has been inadequately greased can often be freed by applying glycerol to the groove between key and barrel at top and bottom of the tap and then heating it with a brush flame and tapping it firmly.

2.2.4.2 PTFE taps   The work of chemists using vacuum apparatus has been made considerably easier by the advent of taps with glass barrels and keys made either partly or wholly out of PTFE. Several varieties of such taps are commercially available and the use of a particular type tends to be dictated by tradition rather than logic. When choosing the type to be used the following points are worth considering: (1) Glass keys with PTFE rings tend to be fragile and can lead to problems. Very thin, unsupported PTFE is not capable of withstanding constant use and will tend to distort. (2) PTFE has a much higher coefficient of thermal expansion than glass and is comparatively soft. Therefore, the larger the area of the PTFE/glass seal the better; small imperfections on the surface will not destroy the seal and it will hold over greater changes of temperature. (3) Where polymeric materials other than PTFE (such as Viton) are used in the construction it is important

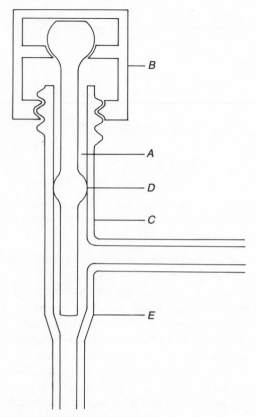

Fig. 2.6. The diagram shows the essential features of the PTFE taps, but it is not a representation of any one firm's model. A is a shaft of PTFE or of some other material covered with PTFE. It is driven by rotating the plastic cap B which engages the screw-thread of the glass moulding C. The seal against the atmosphere is provided by the bulge D. The valve closure at E is the more reliable of the two seals.

that these materials do not come into contact with any chemicals with which they can react either physically (absorption and swelling) or chemically.

PTFE is essentially inert, but very aggressive chemicals, such as SbF_5, may cause deterioration of the sealing ability of PTFE over longer contact times. Therefore, to isolate reservoirs of aggressive chemicals for more than a few days it is better to use a glass break-seal.

PTFE taps usually have two different types of seal as shown in Fig. 2.6. Seal E is produced by a vertical pressure of the key, whereas seal D is the result of a lateral pressure from the body of the key. The latter seal tends to be the weaker and the more temperature-sensitive, and this should be considered when planning the orientation of the tap in the system: the better seal E should always be used to close the more permanent vacuum.

Finally, it should be remembered that the glass parts of PTFE taps are made by moulding and not by blowing, so that they have some internal strain, and therefore the body of the tap must not be heated unevenly when it is fused to another piece of glass.

2.2.4.3. Metal taps. The most frequent use of metal taps on a chemist's vacuum line is at the pumping end where the very large bore metal valves available commercially are both better value for money and more robust than their glass counterparts. Furthermore, at the pumping end it is usually only necessary to have an 'open/shut' valve and for this purpose the large metal valves having an 'O' ring rubber-to-metal seal are ideal. For the chemist planning a new vacuum line the use of metal tubing (especially of the flexible type) as far as the cold traps is also recommended since this leads to a very robust system. The major disadvantage of these large metal valves, and indeed of the metal tubing, concerns the rubber 'O' rings included at each join. Many organic solvents attack Viton, the commonly used material. Also, many inorganic acids can corrode the metal surfaces and render the seal useless. For these reasons a chemist who can afford a pumping system connected with such metal valves would be well advised to take special care to ensure that all vapours are trapped efficiently by the cold traps. The practice of letting the pumps down with dry nitrogen via a vent between the cold traps and the pumps (see Section 2.2.2) prolongs the life of such all-metal systems.

Within the working part of the vacuum system the use of all-metal valves is not common; indeed, since the advent of PTFE taps there are few occasions when such taps are necessary. Nevertheless, it is worthwhile to note that suitable taps do exist for operations involving materials incompatible with both PTFE and grease, or if a good 'static vacuum' (see Section 2.3.1) is required for a long time in circumstances where a glass tap would be inappropriate.

The major disadvantages of metal taps in comparison to PTFE or glass taps are:

(1) They require a lot of space because of the need for glass-to-metal seals (Housekeeper seals) on each side of the tap and even more if small spirals are included. The purpose of these spirals is to absorb any movement which occurs as the tap is manipulated. This is particularly worthwhile for metal valves which often require more force to turn.

(2) They are relatively expensive.

(3) Since they cannot be inspected visually, they are unknown quantities from the point of view of cleanliness and reliability.

Nevertheless, metal valves tend to be more reliable and robust than PTFE or glass taps and therefore last longer, and they are to be recommended when frequent use of a very corrosive material is expected.

Commerical metal valves are available in many sizes and materials and with different geometries and functions (Nupro, Hoke, etc.). A very simple and cheap all-metal valve, devised before the advent of PTFE taps, the BiPl valve, has a closure which consists of a steel ball-bearing fitting into a seat of soft solder, the mobility for the shaft carrying the ball-bearing being provided by a flexible copper diaphragm (Biddulph and Plesch, 1956). The idea of a hard ball making its own seat in a soft base was retained in all subsequent developments and proved its worth in many circumstances and over several decades. The final version, in which mobility is conferred by a metal bellows, and which has not been published previously, is described in Section 2.4. One advantage of this type of valve is that it can be dismantled easily and the seating remade.

A simple all-metal valve built on the same principle, whose seal can be regenerated without dismantling, is also described in Section 2.4, as it has only been published in a thesis (Krummenacher, 1971).

2.2.4.4. Magnetic break-seals and ladders   A very important device much used on the best high vacuum systems is the magnetic break-seal. It consists essentially of a fragile glass membrane separating two parts of a high vacuum system, which can be broken by means of a glass-enclosed iron core inside the high vacuum system, moved from outside by a magnet.

Break-seals come in many different forms and two of these are shown in Fig. 2.7. The form shown in Figure 2.7(a) is easier to make and tends to be more reliable, in that it (almost) always breaks when desired, but the type shown in Figure 2.7(b) has the advantage that, after breaking, the full diameter of the tube is open; this means that liquids or gases flow through the opening more rapidly.

Break-seals are most easily made from thin-walled (ca. 1 mm) glass tubing with an o.d. of ca. 10 mm; Fig. 2.7(b) shows the various stages in the preparation of a break-seal. For a break-seal of the type shown in Figure 2.7(a) the only difference is that instead of the thin glass bubble a short capillary is drawn and curled; the remaining steps in making the break-seal are identical. During the final stage of the preparation of a break-seal, care should be taken not to overheat the glass membrane (or the fine glass capillary), since this will lead to the break-seal becoming too robust. Before sealing a break-seal into an apparatus, it is essential to ensure that the breaker has an unrestricted path of some 10 – 15 cm to the break-seal so that this can be broken easily. The break-seal should be broken from its convex side, since there is less danger of the breaker becoming wedged in the aperture. French workers mostly use stainless steel ball-bearings, and from the concave side of the dome seal, because less force is needed and there is less danger of the ball just bouncing off the dome. If the apparatus is to be separated from the line before opening the break-seal, a simple glass rod with a pointed end can sometimes be used as a breaker, provided that the

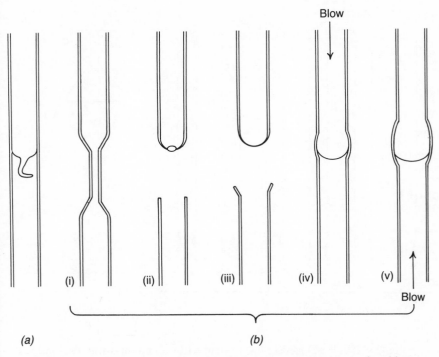

Fig. 2.7. Capillary (*a*) and domed (*b*) break-seals, and the stages of making a domed break-seal. Thin-walled tubing of i.d. 8–15 mm is drawn out as shown in (i), the two parts separated and shaped as in (ii). The small gathering at the centre of the bulb is blown out to a thin dome (iii) and another piece of tubing of the same diameter is flared slightly (iii). The two are fitted together as in (iv) using a needle flame, the blow being from the 'wrong' end. The seal is completed by still working with a needle flame and blowing from the 'right' side (v).

apparatus can be manipulated easily so that the breaker can be 'thrown' against the break-seal. For more delicate apparatus, or if the apparatus is fixed, the breaker will have to be moved by a permanent magnet or by a solenoid. The latter is especially useful if the breaker will have to open the break-seal from below. Although it is usually most convenient to open a break-seal by moving the breaker along the tube in line with the break-seal, the construction shown in Figs 5.5 and 5.11 is often useful, since it reduces the overall length of the apparatus and also allows the breaker to be held out of the tube through which the material is to be transported. This type of breaking device is, of course, unsuitable for the domed break-seal shown in Fig. 2.7(*b*).

In all the examples in this book of the use of break-seals, they are used as once-only devices. However, one can use them as taps for a small number of on–off operations, if the most rigorous conditions must be met. The device

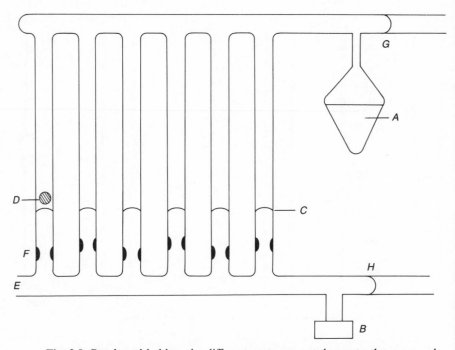

Fig. 2.8. Break-seal ladder; the different parts are not drawn to the same scale. Several portions of known volume of a liquid A are to be distilled into a container or apparatus B. Procedure: One of the break-seals C is broken with the steel ball D and the required volume of A is distilled into a hanging burette (not shown) beyond E. The broken break-seal is sealed off by fusing at its seal-off point F; then this first portion is distilled from the burette into B. For the next operation, D is transferred by a magnet into the next rung of the ladder. When all the break-seals have been used up, a second ladder can be attached at G and H without breaking the vacuum.

for doing this, illustrated in Fig. 2.8, is called a break-seal ladder for obvious reasons, but usually it is used on its side, to make the breaking of the seals easier, as shown. Their mode of operation should be obvious from the diagram.

2.2.4.5. The freeze valve   Another simple 'valve' which is extremely effective and especially useful when one needs to seal off a unit containing a volatile liquid from the vacuum line, is the freeze-valve. This is a glass U-tube of not more than 5 mm i.d. in the duct connecting the two parts which need to be separated. If some of the liquid is frozen into the U so that the tube is blocked, the low pressure side of the U can be sealed off without the necessity of freezing the bulk of the liquid on the other side. This technique can be especially useful when the flask containing the bulk liquid is very large or if

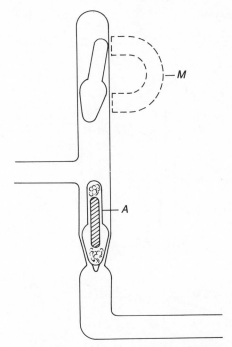

Fig. 2.9. Dry valve. The stopper with an iron core (*A*) and its seat are made preferably from B.10 or B.14 joints. The stopper with an iron core is made in the same way as the magnetic phial-breakers (see Section 1.3.3.1). *M* is a small permanent magnet.

it contains glass parts (such as temperature probe pockets) which would undergo considerable stress if they were included into a bulk of freezing and thawing material; this device is also described in Section 3.1.3. with reference to calibrated break-seals. Its main disadvantage is that the U-tube, which must be kept fairly narrow, is a considerable obstacle to efficient pumping.

2.2.4.6. Dry valves   In some instances it is useful to have a valve at a location in the system across which no significant pressure difference is to be expected. Such a valve can be constructed from a glass socket and a cone with a magnetic core as shown in Fig. 2.9. It has been used, for example, to separate the mixing chamber of a dilatometer from the body and capillary during kinetic measurements, when the distillation of the solvent from one part of the apparatus to another, due to a small difference of vapour pressure, must be prevented. The valve can be kept open by fixing a magnet to the stopper duct with adhesive tape.

2.2.4.7. The Fairbrother separator   A simple device was required for separating the contents of the anode and cathode chambers after an

To vacuum line

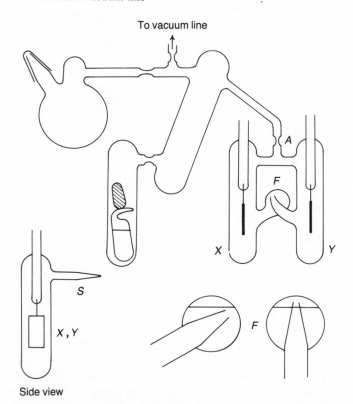

Side view

Fig. 2.10. The Fairbrother separator. For full details see Fairbrother and Scott (1955). Once the electrolytic cell XY has been filled and sealed off at A, the electrolysis is carried out, then the liquid in X and Y is disconnected by tilting the whole appliance, the tips S on X and on Y are broken, and the contents of the cell compartments emptied for analysis.

electrolysis, so that they could be analysed separately. This is shown in Fig. 2.10 (Fairbrother and Scott, 1955). Simply by tilting the apparatus, the liquid contact between the cathode and the anode chambers is broken and the two solutions become separated. It is clear that such an arrangement is not suitable for the long term separation of two solutions of unequal concentrations since the solvent would distill into the chamber with the more concentrated solution. Provided that the separation is not for long and that the vapour pressure of the solvent is relatively low, this method is very useful.

2.2.5. Gauges.

2.2.5.1 General   The measurement of the absolute pressure in a vacuum system is important to vacuum engineers but generally not for chemists. Very few commercially available gauges are suitable for installation directly onto

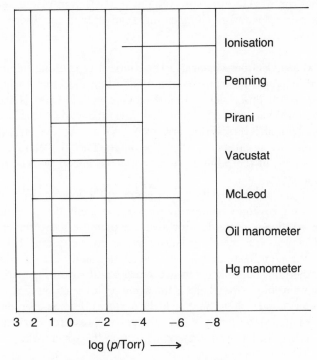

Fig. 2.11. The ranges of various manometers.

the working section of a chemist's vacuum line because of their sensitivity to sudden rises in pressure and their lack of resistance to chemical attack. Indeed, the chemist is seldom interested in accurate pressure measurement (better than an order of magnitude) except when measuring the vapour pressure of a particular compound or mixture. If a gauge is required along the main part of the vacuum line then it must be separated from the system by a tap. A single gauge can be connected to several parts of the system, each connection having its own tap. Sensitive electrical gauges, although convenient, are a luxury and should be treated with the respect that their cost demands.

The uses of gauges on a chemist's vacuum line are to find out whether the system is free of leaks and whether the final vacuum has been achieved. For both these purposes the gauges can be connected to a point between the cold traps and the main line. For systems with a diffusion pump, a gauge connected to a point between the diffusion and backing pumps can be used to indicate when the backing-pressure is low enough to allow the diffusion pump to be switched into the system, but this is very rarely necessary, as the experienced operator can tell from the clicking sound of the rotary pump when this stage is reached.

Fig. 2.11 shows the working pressure ranges of a selection of different gauge types whose principles and advantages and disadvantages are discussed below.

2.2.5.2 Mechanical and electromechanical The simplest type of mechanical gauge is the Bourdon gauge; this consists of a tube of thin sheet metal which has a lentoid cross-section and is bent into an arc. One end of the arc is fixed rigidly and is open to the vacuum to be measured. The other end is closed and attached, via a system of levers and gears, to a pointer. As the pressure changes, the radius of curvature of the tube alters and the change is magnified by movement of the pointer. An aneroid barometer works on a similar principle.

The accuracy of any model of this type of gauge at the low pressure end can be considerably improved by evacuating both inside and outside the tube and measuring in a differential mode. For absolute pressure measurements such mechanical gauges of metal generally cease to be useful below $ca.$ 10^{-1} Torr, but they can be made from chemically resistant and bakeable materials and are therefore suitable for measurements in the working section of a chemist's vacuum line. Nevertheless, the accuracy of such gauges will deteriorate if vapours condense in them or if the surfaces become corroded, and they can be damaged by large, rapid changes in pressure. For these reasons it is advisable to isolate even these gauges from the system when not in use.

Bourdon-type gauges of glass or quartz come in two variations: the spoon gauge and the spiral gauge (Fig. 2.12(a) and (b)). The spoon gauge is approximately the size and shape of a teaspoon, with the access tube to the space between the curved surfaces where the handle of the spoon would be, and a long fibre attached to the tip. This magnifies the movement resulting from the change of curvature of the spoon when the pressure difference between inside and outside of the spoon changes.

The spiral gauge consists of a hollow spiral of very thin glass or quartz fixed at the end at which it is connected to the vacuum line, the other end being closed. Changes in the difference between inside and outside pressure lead to expansion or contraction of the spiral. The resulting movement can be magnified by a mechanical or optical lever.

Both types of Bourdon gauge are most suitable for use with corrosive gases and both can be used most effectively as null-point instruments. Several types of mechanical gauge are available commercially which use electrical capacitance or induction to magnify the mechanical movement of a membrane. Such gauges are easily operated in a differential mode and can be used for measuring pressure differences down to $ca.$ 10^{-2} Torr.

A variety of electromechanical gauges, known as pressure transducers, is available commercially. Most work on the following principle: A flexible metal diaphragm separates two chambers, one of which is connected to the

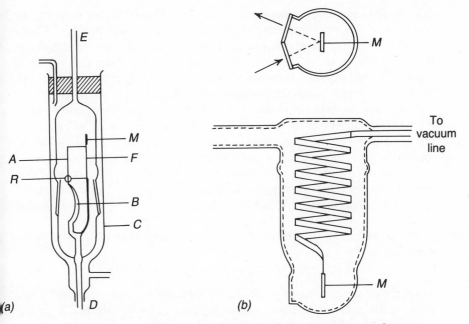

Fig. 2.12. (*a*) Spoon and (*b*) spiral gauges. The device in (*a*) is one of many variants (Foord, 1934). In this modification the movement of the spoon *B* is limited by a safety ring of glass, *R*, which is attached to the capillary leading to the spoon. The movement of the pointer *A* deflects a thin fibre, *F*, to which a mirror *M* is attached. The whole is placed in a water jacket *C*. Access to the space, the vacuum in which is to be measured, is via *D*. The ground joint facilitates the adjustment of the lever system. An advantage of this mode of construction is that when the arm *A* comes into contact with *R*, further increase of pressure still causes a smaller – *ca.* 1/20 – movement of the mirror. This reduction in sensitivity is very useful when balancing the gauge by admitting gas to the envelope through *E*.

(*b*) One of many designs of spiral gauge in elevation and plan. As the difference in pressure between the outside and inside of the flattened glass spiral changes, the mirror *M* rotates and its movement is enlarged by an optical lever.

space in which the pressure is to be measured (the test space), the other to a reference space in which the pressure can be regulated and measured; its value is displayed on an electronic instrument. The distortion of the diaphragm due to a pressure difference across it changes the capacity of a condenser. If the reference space is evacuated to the lowest feasible pressure, a calibration of the capacity of the condenser as a function of the pressure in the test space can provide a direct pressure reading. However, for most purposes it is preferable to use the device as a null-point instrument by adjusting the pressure in the reference space until it balances that in the test space: zero distortion of the diaphragm.

2.2.5.3. Oil and mercury manometers   Mercury U-tube manometers can be readily used down to a pressure of *ca.* 1 Torr with an accuracy of ± 1 Torr and by substituting an oil (such as butyl phthalate, butyl sebacate or Apiezon B) or by tilting the U, measurements down to 10^{-2} Torr can be made accurately; and accurate measurements down to 10^{-4} Torr are possible with a special U-tube manometer, but such measurements are not routine.

There are two types of U-tube manometer; the open-ended manometer is obviously inconvenient for low pressure measurements if an organic liquid of low vapour pressure is used instead of mercury. The small U-tube with one sealed end is the most common and convenient form. The sealed end is closed whilst the tube is under a good vacuum ($p < 10^{-3}$ Torr) and, provided that precautions are taken to preserve the vacuum, the pressure in this reference space (the Torricelli vacuum) can be neglected when calculating the measured pressure.

When oil rather than mercury is used, the conversion of the reading to Torr is given by

$$p/\text{Torr} = (p_r/\text{Torr}) + (h/\text{mm}) \times (\rho_{\text{oil}}/13.6)$$

where the reference pressure p_r can usually be neglected and h is the difference between the levels of the liquid in the two arms of the U-tube and ρ_{oil} is the density of the oil in grams per cubic centimetre.

Since U-tube manometers can be made easily, it is worth noting several tips for their construction:

(1) The U-tube must be uniform in diameter and scrupulously clean to avoid errors due to variations in surface tension. The best final stage for the cleaning process is to heat the evacuated tube to between 280 and 300 °C. A small constriction at the base of the U does not affect the measurements and prevents sudden surges of fluid, which can be destructive.

(2) When mercury is to be used as the fluid, it is best to distil it into the U-tube *in vacuo* to ensure that it is as clean as possible and free from air.

(3) For accurate measurements it is essential to degas the manometer fluid before charging the manometer with it, but if it is introduced by distillation *in vacuo* this is, of course, not necessary.

U-tube manometers are cheap and convenient and can be used in the working part of the line.

Since all organic liquids, and also mercury, dissolve gases and especially the vapours of organic compounds, it is good practice to keep fluid manometers under vacuum and isolated from the system when they are not in use, so that they can discharge any dissolved material into an evacuated space while 'resting'.

2.2.5.4 The McLeod gauge The principle of operation of the McLeod gauge is that a large volume V of gas at low pressure P is compressed into a small volume v contained in a glass capillary. If V and v are known from a calibration and the pressure p in the capillary is known from the difference in height h between the levels of the mercury in it and an evacuated capillary of the same diameter (to eliminate surface tension effects), then

$$P = p \times v/V$$

Since the volume of the compressed gas is $\pi a^2 h$, where a is the radius of the capillary, and p is given by $\rho g h$ (ρ = density of Hg, g = acceleration due to gravity),

$$P = \pi \rho g a^2 h^2 / V$$

Evidently the sensitivity, dh/dP, is the greater, the greater V and the smaller a are. There are many designs of McLeod gauge, but the commonest has a V of between 50 and 200 cm³ and a capillary of *ca.* 1 mm i.d., and the mercury is lowered into the reservoir by suction from a filter pump or other source of vacuum, and raised by admitting air slowly through a fine capillary, both operations being controlled through the same solid-key 'either-or' tap attached to the reservoir. If P is less than *ca.* 10^{-5} Torr, the mecury tends to stick in the capillary; hence the term 'sticking vacuum'.

The device shown in Fig. 2.13, the 'Vacustat' (Edwards High Vacuum Reg. Trade Mark), which is a miniature McLeod gauge, has the advantage that only 8–10 cm³ of mercury is required, and making a measurement is quicker and simpler than with the standard McLeod gauge because it involves merely turning the gauge about its pivot until the level of the mercury reaches the level D. These swivel-type gauges are usually attached to the vacuum line by a greased cone-joint; because they are smaller their accuracy and their range are considerably less than those of the standard McLeod gauge.

McLeod gauges have the advantage of being 'absolute', and they are therefore used for calibrating electrical gauges. Their disadvantages include the use of a large volume of mercury and that they cannot measure the pressure of a condensable gas.

2.2.5.5. Electrical gauges Most electrical gauges require calibration and are 'gas-specific', which means that for different gases the same reading signifies different pressures. On a chemist's vacuum line such gauges are used regularly only to measure the pressure of non-condensable gas between the diffusion pump and the cold-traps or between the pumps, or the final pressure when the main line has been pumped down. As these are not essential measurements, only the more common types of gauge will be described, and only in outline.

(i) *Pirani gauges* At pressures below that at which gas flow changes from molecular to viscous, the thermal conductivity of a gas is approximately

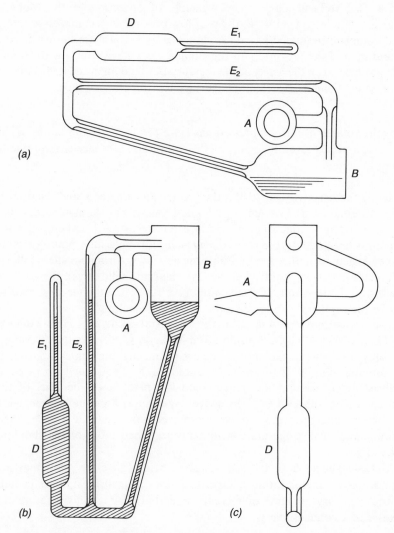

Fig. 2.13. The 'Vacustat' or swivel gauge. This instrument is plugged into an outlet from the vacuum line by means of the horizontal B.14 cone A (see (c)). The normal position of the gauge is, as shown in (a), horizontal, so that the whole of it is evacuated; the mercury is in reservoir B. When a measurement of the pressure is to be made, the instrument is turned into the vertical position, (b), so that the mercury traps the content of bulb D and compresses it into capillary E_1 which has the same diameter as capillary E_2. The pressure of the gas compressed into E_1 is given by the difference in height of the mercury in E_1 and E_2.

proportional to its pressure. The Pirani gauge measures the thermal conductivity which is actually expressed in pressure units. The device consists of an electrically heated filament of platinum or tungsten, both of which have large temperature coefficients of resistance. Either the filament is heated with a constant current and its resistance is measured to indicate the pressure, or the temperature (resistance) of the filament is kept constant and the voltage required to achieve this is measured to indicate the pressure. Both measuring principles involve the use of a Wheatstone bridge circuit and, for the best measurements, an identical filament, sealed in a good ($< 10^{-7}$ Torr) permanent vacuum, is used as a balancing resistance. Commercially available Pirani gauges have short response times and measure pressure between *ca.* 10 and 10^{-4} Torr by means of a range of balancing resistances, but the accuracy of the measurements towards both ends of the scale is usually poor (uncertainly $> \pm 10\%$). The major disadvantage of the Pirani gauge is that for accurate pressure measurements the nature of the gas must be known, since the rate of heat loss from the filament depends on the molar mass of the gas. The measurements are also sensitive to vibrations and, of course, rely on an accurate calibration. Furthermore, Pirani gauges should not be used where they may come into contact with volatile liquids, in particular halogenated organics, for extended periods since this causes the filament to burn out rapidly.

Thermocouple gauges work on a similar principle but have a thermocouple as sensor connected to a heated platinum filament. The e.m.f. of the thermocouple is measured with a galvanometer or potentiometer. Such gauges have a normal working range from 10^{-1} to 10^{-3} Torr, but otherwise have characteristics similar to Pirani gauges.

(ii) *Ionisation gauges*   The ionisation of gas molecules by an accelerated electron beam travelling between a heated filament and a positively charged electrode can be quantified by measuring the current from a third, negatively charged, collecting electrode; the response of these devices is rapid. The measurement of pressure relies on the linear relationship between the current flowing and the gas pressure, provided that this is below *ca.* 10^{-3} Torr. The actual range of linearity depends on the emission current from the heated filament and the accelerating potential. Since the current flowing also depends on the nature of the gas, accurate measurements require calibration for different gases. Ionisation gauges should be degassed thoroughly before use and should not be exposed to pressures greater than *ca.* 10^{-2} Torr. They are not particularly suitable for a chemist's vacuum line since a thorough degassing is required after even short exposures to pressures greater than 10^{-2} Torr before an accurate measurement can be obtained again. Furthermore, although gauges containing a platinum filament are less susceptible to failure due to overheating when accidentally exposed to higher pressures, even these will not survive extended use in contact with organic compounds or other reactive chemicals. Such gauges should be treated with extreme care.

(iii) *The cold cathode gauge*   The working principle of the cold cathode gauge (also known as a Phillips ionisation manometer or a Penning gauge) involves the application of a high voltage between electrodes sealed into a discharge tube. The flow of electrons from the cathode to the anode is forced to follow a spiral course by placing the discharge tube in the field of a strong, permanent magnet. In this way the probability of a gas molecule being ionised by the electrons is increased and a 'magnified' ionisation current can be measured. Pressures as low as 10^{-6} Torr can be measured by such gauges and the response, as for the heated filament gauges, is almost immediate. Excessive currents tend to be induced at pressures greater than 10^{-3} Torr, and the metering electronics of such gauges usually involve a cut-out mechanism to protect their circuits. At very low pressures the relation between current and pressure ceases to be linear. Like other electrical gauges, the cold cathode gauge is gas-specific so that it must be calibrated for accurate work.

2.2.5.6. Choice of gauges   For the general operation of a vacuum system, a vacuum gauge is usually not required, but it may be useful, especially to the less experienced operator. For general monitoring purposes the small U-tube manometers and the 'Vacustat'-type mini-McLeod gauge are adequate.

If vapour pressure measurements are to be an essential part of the work to be undertaken, a cold cathode manometer is probably the best choice, despite the fact that it needs to be calibrated for each molecular species, and its use with mixtures of gases containing two or more species is correspondingly more difficult. If such mixtures are to be investigated, or if the chemicals concerned are corrosive, it is probably most efficient to use a mechanical gauge as a null-point instrument and to measure the pressure by means of a McLeod gauge.

2.3. Building, operating, testing and dismantling the vacuum line

2.3.1 Building and operating
Since it is desirable that as much of the line as possible should be as wide as possible and that the connections at the pumping end should be as short as possible, the pumping assembly and the working line should be constructed of tubing with an i.d. of *ca.* 15–20 mm by a professional glass blower. If this is not possible, the beginner, and even the practised amateur, should resign himself to using *ca.* 10–15 mm i.d. tubing for the pump-end connections and the trunk line, rather than the wider tubing which is more efficient but much more difficult to work. It is better to get a rather 'slow' line up and working

quickly than to struggle for a discouragingly long time to build a more efficient line.

When all the components have been assembled and cleaned, the first stage of building, by professional or amateur, is to clamp the components into what seem the most useful positions, i.e. to make a mock-up. It then generally becomes obvious quite rapidly that this is not as easy as it may seem, because one needs to allow for space to get hands on taps, Dewar vessels around hanging burettes, the hand torch round the back of some necessary gadget, etc. It is well worth while sitting for quite a while in front of the array of components clamped to the supporting grid and thinking of what will need to be done; if there is a colleague or co-worker with whom to discuss the options and procedures, so much the better.

Until a few years ago there would have been no problem as to how the various components were to be joined together. There were only two methods: cone-and-socket junctions or fusion, and for the kind of system which this book is about, as many as possible of the joints would be made by fusion. In recent years, however, a great variety of metal pipework with a great diversity of vacuum-tight junctions has become available, and also some 'push-fit' devices consisting of a pair of O rings, one of PTFE, one of Viton or a similar material, fixed upon a glass tube, to be pushed axially into a precision-moulded wider glass tube.

At the risk of appearing old-fashioned, these various innovations will be ignored in this book because (1) the author has no personal experience of these devices, (2) they do not appear to offer any advantage over all-glass ducts joined by fusion or by traditional cone joints, and (3) they appear to be considerably more expensive than the traditional methods, without offering any clear advantages for the type of system being considered here.

Therefore, fusion being the method of choice, the next stage in the construction of the vacuum system is to 'blow the bits together', according to the plan i.e. to join up the various components, starting at the pump end (see Fig. 2.1). When the backing pump is installed and connected to taps T_1, T_2, and T_3, it is switched on and the progressive evacuation of the space enclosed by T_2 and T_3 is followed by the change in colour of the Tesla discharge (see Table 2.1), and the concurrent change in tone of the backing pump should be noted. If this section is free of leaks, the diffusion pump and ballast volume are connected and the space up to T_4, T_{15}, T_{11} is then tested for leaks, without using the diffusion pump. In this way one progresses, connecting up and testing section by section, as far as T_{12} and T_{13}.

It is very useful to leave each section separated from its neightbours under vacuum over night and to test the state of the vacuum next morning. It may happen that a section gives a good vacuum whilst it is being pumped (*dynamic* vacuum), but that when left shut off, the pressure in it increases steadily – but often quite slowly; such a section is said not to hold a *static* vacuum. However, because it can take several days of pumping to degas oils,

Table 2.1. *The colour of an electrical (Tesla coil) discharge at different pressures; residual gas: air.*

Approximate Pressure $P/$Torr	Colour	Form
5–1	Purple	Thin line
1–10^{-1}	Pink–dark blue	Discharge fills tube close to electrode
10^{-1}–10^{-2}	Dark–pale blue	Discharge extends
10^{-2}–10^{-4}	Light blue–black	throughout system
< 10^{-4}	No Discharge	

In the range 1–10^{-3} Torr hydrocarbons give a grey discharge, halogenated compounds a grey-green discharge.

greases etc., the system should be pumped out for several days and the various sections left under vacuum over night. The vacuum can be monitored by judicious opening of T_{10}.

If there are no leaks, the evacuated traps are immersed slowly in a Dewar vessel full of liquid nitrogen and then the diffusion pump is switched on with T_1 and T_3 open, T_4, T_{11} and T_{15} closed. The residual glow from the Tesla discharge should disappear rapidly from the space between the diffusion pump and T_4. Then T_4 is opened and the traps pumped down, then T_5 is opened and the space bounded by T_6, T_7, T_{10} and T_{11}, and finally the main trunk line T_8–T_{12} and the subsidiary trunk line T_9–T_{13} are connected for 'hard' pumping.

The object of the ballast volume above T_{11} is to provide a backing vacuum for the diffusion pump to pump into, if one does not want to run the backing pump, e.g. over night or over a weekend; it is, of course, essential to make provision for keeping the cooling bath around the traps well filled, and to make sure that there is no continuous supply (leak) of any volatile material entering the pump. This procedure is usually used to clear a system of adsorbed volatile compounds. In order to prepare the backing volume for this, it is pumped out thoroughly with the diffusion pump backed by the backing pump through T_4, T_5 and T_{11}. Then T_3 is closed, T_{11} turned so as to connect the ballast volume to the high pressure side of the diffusion pump, and the backing pump is then shut down. When normal pumping is to be resumed, the ballast volume is connected to the manometer, or tested with the Tesla coil to see how much gas has been accumulated in it. If all is well, there should not be a pressure much above 1 Torr. Normal pumping procedure is then resumed, and the ballast volume is reevacuated at some convenient time. When not in use it should be kept evacuated and not be connected to the line. A small U-tube manometer fused to its neck is often useful.

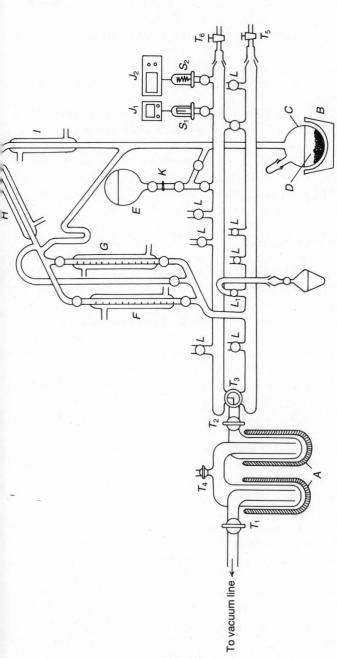

Fig. 2.14. Schematic diagram of a high vacuum system with two working lines, solvent supply systems and pressure gauges. All the taps drawn as circles are PTFE taps. T_1, T_2, T_3 are large-bore glass vacuum taps. T_4, T_5, T_6 can be solid-key glass taps. A are Dewar vessels surrounding twin cold traps. B is a heating mantle in which is the solvent reservoir C containing a solvent and a drying agent D; E is a reservoir for a liquid reagent or solvent. F and G are thermostatted solvent burettes of different sizes, e.g. 10 ml and 50 ml. H and I are condensers. When the burettes need to be filled, the jacket of H is percolated by cooling water, that of I is empty; when the distillation is to be stopped, the jacket of I is filled. S_1 and S_2 are the sensing heads for the pressure gauges J_1 (down to 10^{-3} Torr) and J_2 (down to 10^{-10} Torr). K is a seal-on point or cone joint so that different reservoirs E can be fitted. L are spouts to which the various reagent reservoirs, measuring devices, etc. can be attached as shown at L_1.

When the line is ready for work and the various necessary devices and gadgets have been fitted to it, the mode of starting up is the same as described: each section is evacuated and tested, always starting at the pump end.

A pair of working lines with solvent supply systems and standing, thermostatted, burettes is shown in Fig. 2.14 (Nuyken, Kipnich and Pask, 1981 – slightly modified).

The shutting-down procedure at the end of a day's work will depend on what has been done, what needs to be dismantled or disconnected, whether chemicals need to be returned by distillation to a reservoir from a burette, etc. But, whatever else is done, it is useful to leave as many sections as possible under vacuum, to vent the backing pump to air, and to detach and clean the receivers from the traps; dirty traps, containing residues and with old grease on their joints are a common source of trouble.

2.3.2. Hunting and repairing leaks

One of the most laborious jobs for the vacuum chemist is searching for leaks. Careful design, with a minimum of taps and joints, reduces the probability of their occurrence, but so-called pinholes may appear at any new glass-to-glass or glass-to-metal joint, and if the making of a joint has been careless, pinholes can appear many days later. Pinholes, flaws in the glass which are almost invisible to the naked eye, can only be detected after evacuation of the glass apparatus. Their occurrence is reduced by careful glass blowing, but even professionally prepared glassware sometimes contains these faults. Pinholes are leaks and unless the apparatus is free of these, it will not be satisfactory. The sealing of pinholes is relatively easy, it is the finding of them (and other types of leaks) which can take a very long time and where experience is king.

There are two ways of detecting pinholes. The method of first choice is the Tesla coil, the use of which has been described in Chapter 1. There are, however, some limitations to its use which have also been mentioned earlier (p. 22). The most common of these occurs when a pinhole is suspected at a glass-to-metal junction, or in a region close to any structural feature, such as a clamp or any other piece of metal, that attracts the Tesla spark preferentially, and then a different technique must be used. It works like this: The Tesla discharge is made to produce a discharge glow in a part of the vacuum line as close as possible to the region of the line under suspicion, with that part being connected to the pumps. The resulting discharge will usually have the 'nitrogen purple' colour given by air at low pressure which is entering through the pinhole. One then takes a slow small jet of mains gas – unlit – issuing from a glass capillary and directs it at the suspected region of the vacuum line. As soon as the jet of gas hits the pinhole, the gas drawn into the vacuum line causes the colour of the Tesla discharge to change to a blue-grey. Some workers prefer to use a small twist of cotton wool on the end of a wire, wetted with some halogenated solvent. With this device the colour of the discharge changes to greeny-blue when the halo-compound is drawn into the vacuum line. Leaks along badly sealed cone-joints or inadequately lubricated glass vacuum taps can be detected by both methods.

Most pinholes originate from one or other of two processes: making a joint between two tubes (as mentioned earlier) or sealing off a tube; the latter type can be avoided if, when the seal-off point seems well-formed and is ready for annealing, the region where vestiges of the erstwhile inside of the tube can be detected is heated momentarily with a very hot needle flame to near white heat. Both types of pinhole are best repaired by first warming the whole piece in which they are located, then heating the region of the pinhole with a very hot needle flame and with a glass rod pulling away the flawed glass whilst blowing gently, and then letting the bubble thus formed run back on itself by making use of the differential surface tension effect mentioned earlier.

If a pinhole appears whilst the vacuum line is in use and cannot be opened, a temporary repair can be made with 'black wax' (Apiezon or similar) as follows: The wax is warmed and worked into a short rod about the diameter of a match-stick and about 1 cm long. The region of the pinhole is warmed with a very small flame, the black wax is applied to the pinhole as a spot not more than 2 mm in diameter, and then warmed with a very small flame just sufficiently to flow and wet the glass. Before a pinhole patched up in this way is repaired properly, all traces of the wax must be removed with a solvent, otherwise the charred remains and ash from the wax will spoil the glass and make a proper 'fusion repair' that much more difficult.

Leak hunting is a craft which can only be learnt with practice and it is bedevilled more than most aspects of vacuum work by Murphy's Law (see Chapter 1). Pinholes and cracks usually appear in the most inaccessible places, and for good reason. Just because a region is inaccessible, e.g. the back of a tube-to-tube joint, near a wall or metal support, the glass-blowing will have been difficult and its quality and finish less than perfect.

It should not be thought that faults (pinholes or cracks) can appear only at fused joints. The author once spent three weeks hunting a leak which turned out to be a minute crack in the thick wall of a commercial U-tube manometer, and (of course) on the side towards the plank to which the manometer tube was fastened, invisible from the front.

2.3.3. Dismantling
When it is necessary to dismantle a vacuum line, as much care is needed as during its construction, but the considerations are different. This is because the well-used line will probably have attached to it various reservoirs of solvents and liquid reagents and containers with drying agents (Na films, liquid Na + K alloy, molecular sieves, etc.), and there will be condensers

attached to water supply and drainage pipes, electrical connections to heaters, gauges, thermocouples, etc. So the dismantling should be done in several stages:

(a) All electrical leads must be detached and made safe.
(b) Connections to other external supplies, especially water, are disconnected.
(c) The vacuum in the main and any subsidiary lines is let down, but any vessels containing reagents which react with the constituents of air are kept evacuated.
(d) All devices and containers attached by ball- or cone-joints are detached. Any containers containing highly reactive compounds, e.g. metal alkyls, alkali metals or their alloys, etc. are stored securely until they can be made safe in the appropriate conventional manner.
(e) The sections which are fused together are separated by cutting the glass.

Since the mobility which is required when cutting a length of glass tubing in the normal way is lacking, a different technique, called 'hot-spotting', must be used. This is how it should be done: With a good, sharp glass knife make a clean, cut 2–5 mm long at right angles to the tube to be cut, if possible at a place that has not been worked. It is useful to practise the cutting and the subsequent procedure on unattached pieces; the difference between making a shallow ineffective scratch and a clearly cut notch which 'opens' the structure of the glass will soon be perceived by feel and sound. The notch is then wetted: saliva, because of its low surface tension, is better than water. Next, the hand torch with a very hot small flame is held ca. 5 cm above the notched tube with the flame parallel to the tube, the holding hand being steadied against some suitable support. With the other hand a glass rod is introduced into the flame and when the tip is bright red or hotter, it is applied swiftly to the notch with light pressure. If all goes well a sharp cracking sound will signal that the tube has cracked cleanly all round. Because of the nature of the eye–hand co-ordination, this operation is done most efficiently if the eye is fixed on the notch where the hot tip is to land as swiftly as possible, rather than on the rod in the flame. If the tube does not crack, it should be left to cool, a new notch cut, and the procedure repeated. If the crack does not complete the circle, thus separating the two previously joined parts completely, it can be made to grow to completion by tapping the tube gently with e.g. a light screw-driver or knife near the incomplete crack and, if that does not work, by hot-spotting just behind one end of the crack.

No vessel containing a highly reactive compound must be opened by hot-spotting in the open laboratory, but any such must be made safe in some other way. The simplest method is to detach it from the line and then smash it, with appropriate precautions, in a safe destruction area.

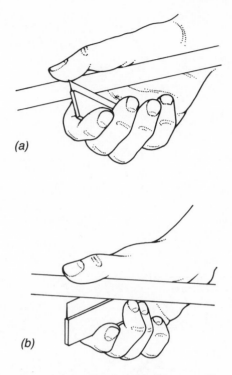

(a)

(b)

Fig. 2.15. The correct (a) and wrong (b) positions of the thumb when cutting a fixed glass tube.

Safety Note   Making the incision for hot-spotting is best done by a special application of the glass-knife to the tube which, being difficult to describe, is illustrated in Fig. 2.15(a). Note that it is the tip of the thumb that supports the glass tube, so that if it should shatter, there is at worst a flesh wound. If the joint of the thumb were used (Fig. 2.14(b)) a tendon would be at risk!

2.4. The PlEgli and Krummenacher valves

The valve to be described here was developed by the author with a Swiss engineer, Mr H. Egli, from the BiPl valve (p. 44). Like its predecessor, it is entirely of metal and free of all kinds of organic sealants, such as PTFE or Neoprene. It will hold a high vacuum against atmospheric or elevated pressure, on either side of the seal. This seal consists of a stainless steel ball *A* which makes its own hollow in the soft-metal (solder) coating covering a circular orifice *B* in a brass casting (Fig. 2.16(a)). The movement for withdrawing the steel ball from its soft-metal seating is provided by a copper

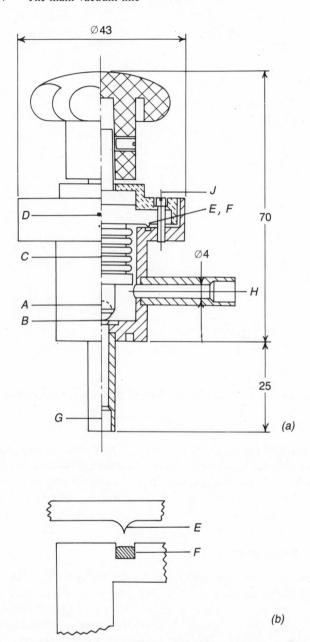

Fig. 2.16. The PlEgli valve. *A* stainless steel ball bearing, *B* orifice to be closed by *A*, *C* copper bellows, *D* cover-plate, *E* sharp annular ridge, *F* channel full of solder, *G*, *H* sockets for Housekeeper seals, *J* screws securing *D* to body of valve.

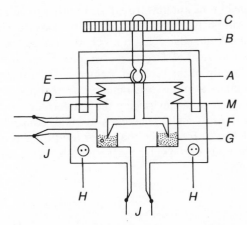

Fig. 2.17. The Krummenacher valve. *A* yoke, *B* spindle, *C* handle, *D* metal bellows, *E* ball-and-socket joint, *F* annular knife-edge, *G* annular trough filled with soft metal, *H* cylindrical heating elements fitting into cavities in the metal body *M*, *J* sockets for Housekeeper seals.

bellows *C*. This is soldered to a cover-plate *D* whose vacuum-tight attachment to the body of the valve is achieved by a special device shown in Fig. 2.16(*b*). The cover-plate carries a sharp, shallow circular ridge *E* which, when the cover-plate is screwed down on to the body of the valve by means of the three screws *J*, bites into solder contained in a circular channel *F*.

If the valve needs to be dismantled by unscrewing the screws *J* then, before reassembling it, one warms the body gently to melt the metal in the channel so that it reforms a smooth surface into which the ridge on the cover-plate can then bite again to make a fresh seal. This device requires much less pressure than lead, indium or gold gaskets and was found much more convenient. Like any other all-metal valve, the PlEgli valve can be connected to a glass vacuum line by metal-to-glass (Housekeeper) seals soldered into the sockets *G* and *H*.

The advantages of the PlEgli over the BiPl valve and PTFE taps include the possibility of dismantling it without cutting it from the vacuum line, its resistance to aggressive compounds such as PF_5 and $TiCl_4$, and insensitivity to temperature changes.

The Krummenacher valve (Krummenacher, 1971) shown in Fig. 2.17 has the further advantages over the PlEgli valve that it has a wider average diameter, i.e. offers less resistance to gas flow, and that its sealing can be regenerated without dismantling it; it is, however, larger and heavier than many other metal valves. Its construction and operation are as follows:

A yoke *A* attached to a cylindrical body *M* carries a threaded spindle *B* operated by a handle *C*. It is attached to a metal bellows *D* by a ball-and-socket joint *E*. By this means the annular knife-edge *F* can be thrust into the tin or other soft metal contained in the annular trough G. In order to achieve a gradual equalisation of pressure between the two sides of the closure when the valve is opened, the annular knife-edge F has a notch in the shape of an inverted V; in this way the withdrawal of F from its bed of soft metal first establishes a small leak when the tip of the inverted V emerges from the soft metal bed. This bed can be remade as often as is required by energising the two cylindrical heating elements H and thus raising the temperature of the valve to the melting point of the soft metal. The glass-to-metal connecting tubes are soldered into the sockets J. Evidently, the solder used for this and for soldering the bellows to the body of the valve must have a higher melting point than the metal in the through G.

References

R. H. Biddulph and P. H. Plesch, *Chem. and Ind.*, 569 (1956).

M. Cole, (*a*) *J. Vac. Sci. Technol.*, **A5**, 4 (1987); (*b*) Private Communication. More information available from M. Cole, Genevac Ltd., Alpha Works, White House Road, Ipswich, IP1 5LU, UK.

F. Fairbrother and N. Scott, *J. Chem. Soc.*, 452 (1955).

S. G. Foord, *J. Sci. Instruments*, **11**, 126 (1934).

B. Krummenacher, Thesis, ETH, Zurich, 1971.

O. Nuyken, S. Kipnich and S. D. Pask, *GIT Fachz. f.d. Laboratorium*, **25**, 461 (1981).

3 Appliances and procedures

3.1. Measuring devices and related manipulations

3.1.1. Burettes

Two orientations are possible for a burette attached to a vacuum line; the 'standing' burette and the 'hanging' burette. A standing burette has its regulating tap below the scale and utilises gravity and/or vapour pressure of the liquid in the burette to cause the liquid to flow from it (Figs. 2.14 and 3.1). The hanging burette has its regulating tap above the scale and the liquid is distilled from it into the recipient vessel, e.g. a reactor (see Fig. 5.9.). It follows that pure liquids can be metered with both types, whereas solutions can only be dispensed from standing burettes. Although in principle compounds having very low vapour pressures at room temperature, such as nitrobenzene or antimony pentafluoride, can be distilled from hanging

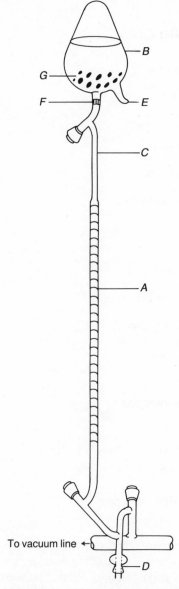

Fig. 3.1. Standing burette A with reservoir B. After B has been filled by
distillation on a vacuum line, it is sealed off from this at E and sealed onto A
at C. A sintered filter F prevents the drying agent G fouling the PTFE tap.
After the recipient for the liquid from B, e.g. a reaction vessel, has been
attached at D, it is evacuated, together with A, A is then filled from B and the
required volume discharged through D. The internal seal there ensures that the
liquid does not come into contact with D. Of course, the joint at D can be
replaced by a fused-on seal.

burettes, this is usually slow and awkward. Lastly, standing burettes are generally more convenient for volumes larger than *ca*. 20 ml.

Before installing the appropriate type of burette, it must be known what volume of material will be involved and to what accuracy the volume is to be measured. For increasing accuracy the burette must be thermostatted. However, before going to the trouble of thermostatting a burette, it is worth considering the accuracy of the burette as a function of the volume to be measured and the error induced by a given temperature change. Consider a 10 ml burette with scale markings every 0.1 ml. The nominal accuracy is not better than ± 0.05 ml which, if 5 ml is to be measured out, is $\pm 1\%$. For many common solvents at 20 °C the change in density with temperature is *ca*. 0.1%/degree, and the errors in many analytical techniques are considerably greater than that. Therefore, little would be gained from thermostatting the burette to better than ± 5 K; indeed, it is rarely worth thermostatting a burette to better than ± 1 K. Hanging burettes can be thermostatted by a simple water bath, but one must remember that as the liquid is distilled, it will cool and therefore several readings should be made to ensure that temperature equilibration with the bath has been established. Often an ice bath is most convenient, especially as for most liquids the density of 0 °C is known accurately. For a standing burette, a glass jacket is constructed through which the thermostatting fluid can flow, similar to a Liebig condenser (Fig. 2.14). This jacket can be most easily constructed from a glass tube with an i.d. *ca*. 1 cm larger than that of the burette, both ends being sealed with a soft rubber O ring. Burettes can be made most easily from sections of standard burettes made of borosilicate glass. For standing burettes it is useful to calibrate the dead volume between the lowest scale-mark and the tap so that the burette can be emptied completely.

3.1.1.1. Standing burettes Standing burettes are frequently used for metering solvents into reaction vessels and a typical arrangement, shown in Fig. 3.1, has been used for metering nitrobenzene (KPG – see references); the device illustrates several useful points. The solvent reservoir is above the burette since distilling nitrobenzene is a very laborious operation. The nitrobenzene is stored over a drying agent, hence the need for a sintered glass filter. It is an advantage that the solvent is drawn through the drying agent on its way to the burette. Finally, the orientation of the PTFE taps is important; the less reliable seals (see Section 2.2.4.2) enclose the burette where the vacuum is required for the shortest time.

A special arrangement of taps and ducts below a standing burette is shown in Fig. 3.2 and has proved useful for metering freshly distilled styrene into a reactor (KPG). The device allows any excess in the burette to be returned to the reservoir at the end of the metering and is useful when fresh distillation immediately before the reaction is desirable, as is the case for thermally polymerisable monomers. It also prevents residual monomer from polymerising in the ducts.

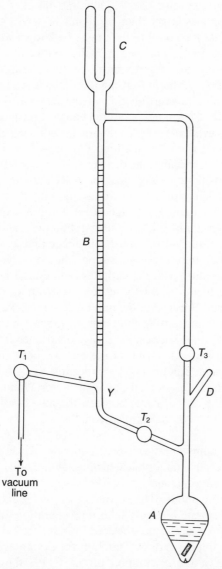

Fig. 3.2. Monomer burette for photopolymerisable monomers. *A* is a reservoir containing monomer over a drying agent, e.g. CaH_2, with magnetic stirrer. The monomer was run into *A* through *D* which was then sealed. *C* is a cold finger to be filled with a mush at just above the freezing point of the liquid in *A*, so that the condensate drips into the burette *B*. Any excess is returned to *A* via T_2 which, like T_1 and T_3, should be a PTFE tap. The rig should be covered in black cloth up to T_2 and T_3. The latter is essential, because in its absence the monomer in *A* will polymerise on all the glass surfaces, even if *A* is kept dark. The reactor in which the monomer is required, or any phials to be filled, are attached below T_1.

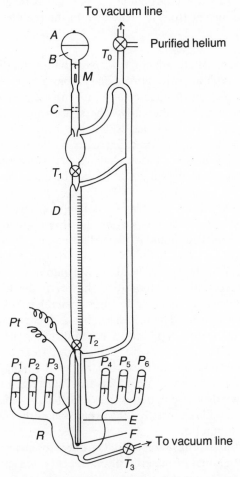

To vacuum line

Purified helium

Fig. 3.3. Pressure equalisation burette in a device for potentiometric titration. For full details of operation see original publication. A break-seal ampoule containing the titrating solution B which reacts with the contents of break-seal ampoules P in reactor R, C sintered filter, D 10 ml burette, E 2 mm capillary, F capillary tip, Pt wires from electrodes leading to potentiometer. T_1, T_2, T_3 PTFE taps, T_0 three-way glass vacuum tap.

When the pressure in the reactor is greater than in the burette, one has two options. One can thermostat the burette at a higher temperature than the reactor but this may be undesirable or not feasible. An alternative is to introduce an inert gas into the reactor and to incorporate a pressure-equalising by-pass in the design. Such a device has been described for use in conjunction with potentiometric measurements under rigorously dry conditions (Jagur-Grodzinski, Feld, Yang and Szwarc, 1965) and is shown in

Fig. 3.3. If a pressure by-pass is used without either an inert-gas blanket or a PTFE tap, distillation of the solvent from the reactor into the burette (or *vice versa*) may occur.

Standing burettes with reservoirs   There are three ways of filling standing burettes: by distillation or by syphon or by gravity feed from a reservoir. The distillation method (Fig. 3.2) is, of course, only available for single compounds; solutions must be handled differently. The syphon method is hardly ever needed and the method of choice is gravity feed with the arrangement shown in Fig. 3.1. The vapour space increases as the solution is used up and so the composition of the solution changes as it is used up; but the resulting change in concentration will generally not be important.

3.1.1.2. Hanging burettes   A typical arrangement for a hanging burette is shown in Fig. 5.9. The splash-heads on the burettes B_1 and B_2 are useful to stop the liquid from flowing into the line should it 'bump' during distillation, which is a particularly acute problem when distilling from a narrow tube such as a hanging burette. The form of the splash-head is not important and traditional forms vary from laboratory to laboratory; however, the form shown in the figure is relatively easy to make and causes a minimal loss of pumping speed. Some forms of splash-head involve sharp bends or capillaries and these should not be used in vacuum apparatus. When distilling into a hanging burette, care should be taken if the material is allowed to freeze. The frozen material should not simply be left to thaw, but rather the hanging burette should be warmed gently and evenly with a hair-drier, starting at the top and working slowly downwards as the material thaws. If this procedure is not followed, there is a considerable chance that the expansion of the solid before it thaws will fracture the burette.

Both hanging and standing burettes have some disadvantages for metering smaller quantities (up to a few grams) of material into a reaction vessel:

(1) The addition requires at least a few seconds to make.
(2) It is tricky to ensure that a precisely known amount is added.
(3) Only liquids or solutions can be handled, not solids.
(4) On an apparatus separated from the vacuum line a burette is very awkward and therefore fragile.
(5) Burettes can restrict the flexibility of design of both apparatus and experiment.

Therefore, in order to meter small quantities of materials *in vacuo*, the use of glass phials or calibrated break-seals, described in Sections 3.1.2 and 3.1.3. respectively, is recommended. Even smaller quantities of moderately volatile materials can be dosed by means of calibrated vapour-dosing bulbs (see Section 3.1.5).

3.1.2. Glass phials

3.1.2.1. Making and testing phials Sealing a small quantity of material into a glass phial and subsequently breaking the phial into a reaction mixture is a method which is used not only with vacuum systems, but is also employed in bomb-calorimetry for the 'instantaneous' addition of reactants. If the material is introduced into the phial *in vacuo*, a ground-glass joint may be joined to the phial (see Fig. 5.4) so that all the glass is still available after sealing off the phial and therefore the weight of the contents can be calculated. However, the use of ground-glass joints is not recommended since, apart from the joints being a possible source of leaks and unknown volatiles (from the grease used), there is a possibility, if liquids are being handled, of the grease being washed into the phial. For this reason it is preferable to have the socket on the apparatus or line and the cone attached to the phial. Furthermore, ground-glass joints are relatively bulky and restrict the number of phials that can be filled in one operation, even if B.10 joints are used.

However, if the joint is dispensed with, it is no longer possible to weigh the glass before and after filling the phial and therefore the weight of the contents becomes an unknown quantity. For many years this problem was solved by collecting the glass fragments for weighing after the reaction. Obviously, this method is not very reliable and it does not solve the problem if the contents of more than one phial are added to a single reaction mixture.

A very elegant solution to this problem is the 'mid-point method' (Rutherford, 1962), and its subsequent developments (Pask, Plesch and DiMaina, 1981). The method is suitable for weighing small quantities (*ca.* 0.1–4.0 g) of liquids, solutions or solids in phials filled under vacuum and it is described in detail below, because although now quite old, it does not seem to be very widely known.

The principle of the mid-point method is as follows (see Fig. 3.4): A length of glass tube is cut and a scratch mark made with a glass cutter at its mid-point. A phial is then blown at one end of the piece of tubing and, after weighing, the tube is sealed to a filling device. After the phial has been filled, it is sealed off from the apparatus ready for subsequent use. The calculation of the weight of the material contained in a phial using the mid-point method is also demonstrated by Figure 3.4., where W_g is the weight of the whole length of the glass tube and the empty glass phial blown from it before it is sealed onto the filling apparatus, W_t is the weight of the glass phial + its contents + the remainder of that half of the original length of tube from which the phial was blown. This piece of tube is simply broken off at the mid-point after the phial has been sealed off from the vacuum apparatus. The weight of the contents W_c is then given by:

$$W_c = W_t - W_g/2 \qquad (3.1)$$

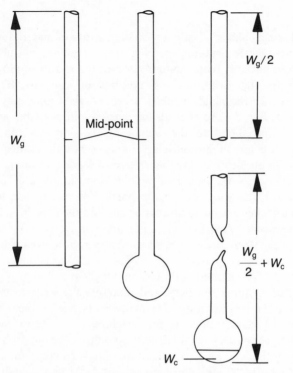

Fig. 3.4. The mid-point method for determining the content of a breakable phial. For explanation see text.

The accuracy of the method is very dependent on the care and reproducibility with which the lengths of tubing are cut and the breaks are made, but under optimum conditions has been estimated as $\pm 1.7 \times 10^{-3}$ g.

In order to improve the reproducibility with which lengths of tube can be cut and marked at their mid-point, a tube cutter has been designed. Such a cutter is not available commercially, but it can be made in any well-equipped metal workshop from the published description (Pask, Plesch and DiMaina, 1981).

(a) For most purposes, phials of 8–16 mm diameter, which have a volume of $ca.$ 1.0 cm³, are suitable and these can be blown easily from tubing with an o.d. of $ca.$ 4.0 mm and $ca.$ 1 mm wall thickness. Experience has shown that the distance between the phials which allows adequate room for sealing off, but which avoids making the apparatus for filling the phials too unwieldy, is 3–4 cm.

(b) The cutting of glass tubes is a laborious task and generally at the start of the operation a high failure rate can be expected. The most common mishaps are either that the tube is scored too deeply and it then breaks at the mid-point as well as at the end, or it is not scored deeply enough and breaks unevenly. Another common mishap is that the glass tube is not rotated absolutely in line with the cutter groove; this too causes the tube to break unevenly.

(c) The cut tubes must be cleaned before use, and this is best done by soaking them successively in chromic acid, water, ammonium hydroxide and finally water again. The tubes are then dried *in vacuo* and kept free from dust until the phials are to be blown. The cutting and cleaning processes can be made more time-effective by making batches of *ca.* 100 lengths of tube.

(d) The clean glass tube requires no further preparation for phial blowing. The flame used should be just hot enough to bring the glass slowly to a workable temperature, and the glass tube is slowly rotated in the flame until the end is sealed and *ca.* 5 mm of glass are at the working temperature before the phial is blown. If the glass is too hot the phial walls will be of uneven thickness and too fragile. When blowing phials it will be found that those which are robust enough to withstand a vacuum and the usual manipulations, but are also fragile enough to be broken easily when desired, give a particular sound (close to middle C) immediately one ceases to blow. Finally, it is worth noting that perfectly spherical phials are more difficult to break than oval phials.

Although the mid-point method requires practice, it is not as time-consuming as it may appear at first sight, since it is possible to fill batches of up to 20 phials in a single operation. This has the distinct advantage that the impurities are the same, at least from this source, for a particular set of experiments in which these phials are used.

As explained on p. 22, thin glass phials cannot be tested for leaks directly with a Tesla coil, and must therefore be tested by the other method (Section 2.3.2) if they are suspected of being leaky. If a phial is leaky, it must be fused off and discarded. Because of the deleterious effect of a narrow tube on the pumping speed, the apparatus for filling phials should be joined to the vacuum line as close as possible to the pumps.

3.1.2.2. Filling the phials.   A batch of phials can be filled with a liquid or a solution by means of the apparatus shown in Fig. 3.5, which is known as a tipping device (KPG). The phial R (see note (ii) below) of material from which the solution is to be made is sealed into the mixing chamber A with a magnetic breaker M, and the apparatus is sealed on to the line at B. After evacuating the apparatus, the appropriate amount of solvent (see note (iii)) is distilled into the mixing chamber A from a reservoir attached to the vacuum line, and the apparatus is then sealed off at the constriction C. The phial R is then broken (see note (iv)) and, after mixing, the apparatus is

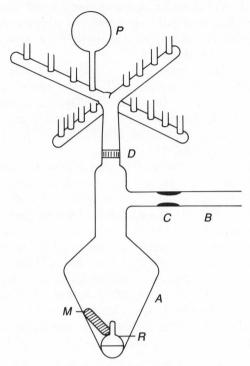

Fig. 3.5. Tipping device for making a solution and distributing it into phials *P*.
See text.

tipped to allow the solution to be filtered through the sinter *D* into the phials
P. If the solvent has a low vapour pressure at room temperature, the solution
can be filtered more rapidly by warming the solution or cooling the phials
before transferring the solution from the mixing chamber to the phials.
Having distributed the solution into the phials (see note (v)) these are cooled
and sealed off (see note (vi)).

The following points for designing and manipulating a tipping device are
worth noting:

(i) The glass sinter *D* serves to retain any solid which has not dissolved
and the fragments of the phial. For all purposes except UV spectroscopy (for
which a grade 3 or 4 sinter should be used), a grade 1 sinter is adequate. The
form of that part of the apparatus to which the phials are attached is
determined by the number of phials to be prepared and the shape of the
Dewar vessel which will hold the cooling liquid to be used when the phials
are sealed off. A form that is relatively easy to make and allows 16–20 phials
to be prepared is a simple cross. The phials should be attached with sufficient
space between them (*ca.* 3–4 cm) to allow them to be sealed off easily. The
seal is best made with a small flame, approximately as wide as the tubing

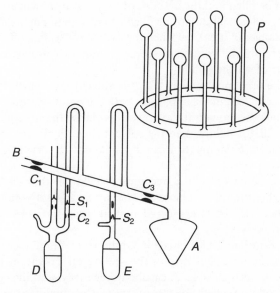

Fig. 3.6. Annular tipping device. See text.

from which the phials have been blown. A small nipple is blown in the position where the phial is to be attached, a hole is blown in the summit of the nipple and then both the rim of the hole and the open end of the phial stem are heated and pushed together. The joint can then be finished in the usual way. Before attaching the apparatus to the vacuum line one must ascertain that all the phials can be cooled simultaneously, (i.e. that the diameter of the assembly is less than that of an available Dewar vessel).

An alternative arrangement for charging breakable phials involves an annular tipping device (Krummenacher, 1971), see Fig. 3.6. The device is connected to the vacuum line at B and sealed at C after evacuation. Ampoule D contains the pure substance to be diluted or a concentrated stock solution. The break-seal S_1 is broken and part or all of the contents of D are poured into the mixing vessel A, and D (being sealed off at C_2) can be used again by way of its second break-seal. Then the solvent in E is transferred to A, which is sealed off at C_3, and its contents are distributed into the phials P by rotating the ring. The phials are then equilibrated and sealed off as described above.

(ii) If materials are to be dissolved which have not been prepared *in vacuo*, it is better either to seal these into a phial (which can be broken before pumping out the apparatus) or to place them in a small glass boat. Remember that when the apparatus is being fused to the vacuum line, there will be an open connection between the mouth and the material.

(iii) The amount of solvent is determined by the concentration desired, the number of phials to be filled and the amount of material from which the solution is to be made. A phial should not be filled to more than one half its volume because of the danger of breakage when it is frozen for sealing off (see also note (vii)) and therefore the volume of solvent required is usually one half of the total volume of all the phials.

(iv) If the phial to be broken is heavier than the solvent into which it is to be broken, it should be placed in the apparatus with the point up, towards the breaker, as this avoids the danger of the apparatus rather than the phial being fractured. If the phial is lighter than the solvent it can be frozen into the surface of the solvent for breaking, or it can be broken with very little or no solvent in the breaking vessel.

(v) Before making a batch of phials, one must decide whether all the phials should have roughly the same volume or whether a range of volumes is required, and it should also be remembered that at least one phial will normally be needed for analysis.

(vi) When working with batches of phials it is important to number the phials clearly and to make sure that during the sealing-off operation the numbered phials are easily recognisable. A cardboard egg tray with the egg-holders suitably numbered is excellent for this purpose. The phial itself must *not* be marked with a grease pencil or in any other way.

(vii) Having distributed the solution into the phials, the stems should be washed to ensure that no solution remains there by stroking them with cotton wool soaked in liquid nitrogen. The phials should then be cooled to about 60–100 K below the normal boiling point of the solvent, and then allowed to warm up slowly (over several hours) by 20–40 K. This final temperature must be such that the vapour pressure of the solvent is so low that only negligible amounts of decomposition products are formed during the sealing off process. At that stage the phials can be sealed off. It is important that the temperature in the cooling bath is kept uniform to prevent distillation of solvent from the warmer to the colder phials. A useful cold bath for this job is a methylene dichloride mush. This has the advantage that methylene dichloride is non-flammable and the temperature of the mixture can be adjusted easily between −93 °C and room temperature. The phials are first sealed off about 1 cm below the point where they are joined to the assembly and set carefully aside, for example in the appropriate compartment of the numbered egg tray. During this sealing operation one should always work with the flame directed from the assembly outward towards the phial. As soon as the narrow tube is sealed there is no need to work on the assembly side of the seal. When all the phials have been sealed off, they are then separated from their stems. This is best achieved by freezing the contents of the phials in liquid nitrogen so that the stems can be sealed off as close as possible to the phial (*ca.* 0.5 cm). If the phial has been filled to more than half its volume the danger of breakage during the freeze–thaw process can be minimised by submerging the phial in liquid nitrogen with its stem at an

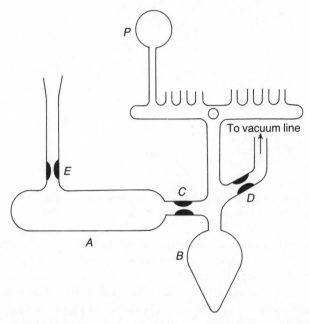

To vacuum line

Fig. 3.7. Tipping device arranged for sublimation. The solid is introduced into *A* through *E* which is then sealed off. The assembly is evacuated and sealed off at *D*. The solid in *A* is then warmed gently so that it sublimes beyond *C*, and by cooling the flask *B* and by dislodging any crystals formed in the parts above it, most of the solid can be collected in *B*. After sealing off at *C* the solid can be distributed into the phials *P* by pouring, if it is a free running powder, and/or by sublimation.

angle. The phial is then rotated so that as the contents freeze they leave a hollow at their centre.

(viii) The numbered egg tray, mentioned above, is an excellent place to store the prepared phials for later use.

Filling phials with solids requires that the solid can either be sublimed or that it is a free flowing, fine powder, because otherwise pieces will stick to the glass or block the narrow tubes and make sealing off impossible.

When subliming material into phials, great care must be taken never to apply heat directly to the phials since they are easily broken. A simple apparatus for subliming a solid from a reservoir into phials is shown in Figure 3.7 (KPG).

A final tip for work with materials, such as aluminium bromide, whose density is relatively large, is not to overfill the phials since they will become very fragile if they contain more than *ca.* 3.0 g of material.

When phials of a solid are required and the material cannot be sublimed or distilled, it can still be brought into the phial as a concentrated solution

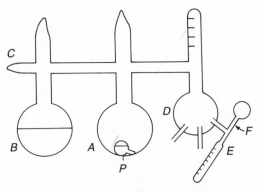

Fig. 3.8. Arrangement for filling phials with a known volume of a solution of known concentration. For operation details see text.

and the solvent can then be removed by distillation and finally by pumping (Jones and Plesch, 1979). The major difficulty with this technique is to ensure that the narrow stems of the phials are free of material so that no problems are encountered when sealing off the phials. The trick involved is to distil most of the solvent from the filled phials very slowly and then to wash the stems by stroking them with a small ball of cotton wool soaked in liquid nitrogen; this causes the remaining solvent to condense in the area being cooled, so that any remnants of solute are washed down into the phial. This operation requires care and patience but it is important that no traces of material are left near the seal-off points if the phials are to stay free of decomposition products. When phials are filled by this method they should not be filled to more than about one quarter of their capacity, otherwise the distillation of the solvent without 'bumping' becomes very difficult.

A tipping device can also be used to follow the course of a reaction, if the reaction mixture is distributed into the phials which are then sealed off and subsequently broken and the contents analysed at intervals (Jones and Plesch, 1979 and Section 5.3.4).

Another method for filling phials, for which the apparatus is shown in Fig. 3.8, has been described by Stannett, *et al.* (1976). A phial P containing the pure solute is placed in the central flask A and, after evacuation and introduction of the solvent from B, the apparatus is sealed off from the vacuum line at C. The solute phial is then broken and the solution tipped into the flask D whence it can be distributed into the side arms E, each consisting of a small burette attached to a phial. The burettes are then cooled, the side arms sealed off, and then the individual side arms can be inverted to fill the phials with the known volume of solution. Lastly, the phials are separated from the burettes by sealing them off at F.

One advantage of this method compared with a tipping device and mid-point phials, may be that a better defined range of volumes can be achieved for a batch of phials, and the method would also be suitable for preparing a batch of larger (say 5–10 ml) known volumes from a single solution. However, volumes greater than *ca.* 3.0 ml are too great to be handled in phials and are better stored behind break-seals; these are described in the next section.

3.1.3. Calibrated vessels with glass break-seals

An alternative to the use of small spherical phials, which often has advantages in terms of the flexibility of the apparatus, is the use of calibrated vessels or ampoules with break-seals. They consist usually of a tube 3–6 mm i.d. and 20–70 mm long with a calibration mark and carrying a fragile glass membrane which can be broken when desired, so as to admit the contents of the ampoule to a mixing vessel or reaction vessel. The construction of different types of magnetic break-seals has been described in Section 2.2.4.4.

Calibrated vessels with break-seals can be made simply by sealing a length of narrow bore (*ca.* 2 mm i.d.) tubing at an appropriate distance from the concave side of the break-seal. The break-seal is then filled with a known volume of liquid so that the surface of the liquid is at least 2 cm above the join of the narrow bore tube with the break-seal. A 2 cm length of narrow bore tube above and below the surface of the known volume allows reasonable flexibility when the break-seal is subsequently filled under vacuum. A scratch is then made on the glass to mark the surface of the known volume. After a calibrated break-seal has been fused to a vacuum apparatus, evacuated and filled with the required liquid, it can be sealed off from the apparatus without freezing the whole unit (not to be recommended in any case) by freezing a 'plug' of solvent in the narrow bore tube, at least 2 cm away from the seal-off point, by holding a piece of cotton wool soaked in liquid nitrogen on to the tube for a few minutes. If such a device cannot be filled exactly to the mark, the volume contained in it is found by measuring the distance of the meniscus from the calibration mark and calculating the volume by means of the calibration of the narrow tube (in millilitres per centimetre). (See Fig. 3.9).

Calibrated break-seals can be used for larger volumes ($> ca.$ 2.0 cm^3) of both pure liquids and solutions with great accuracy. The accuracy depends on the care taken in calibrating the volume of the break-seal, since the error arising from errors in measuring the distance between the calibration mark and the meniscus is negligibly small. For example, a tube with an i.d. of 2 mm contains 3×10^{-2} cm^3 per cm of length, so that an error of 1 mm produces an error of 3×10^{-3} cm^3.

This section would not be complete without a reference to what is probably the best-known form and the most common usage of break-seals,

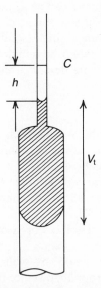

Fig. 3.9. Calibrated break-seal ampoule. If V_c is the volume to the calibration mark C, the content as shown is $V_t = V_c - \pi r^2 h$ where r is the radius of the narrow tube.

namely those used by most workers on anionic polymerisations. These were devised by Szwarc's group at Syracuse and developed by the groups of Bywater, Sigwalt, G. V. Schulz and many others. The basic devices have been described by Szwarc (1968). As they are also to be found in other books and numerous papers we can refrain from describing them here; examples are shown in Figs. 5.2 and 5.8

3.1.4. Phial filler for large volumes
Krummenacher has described an ingenious device suitable for metering relatively large volumes of pure liquids or solutions into break-seal ampoules; it is illustrated in Fig. 3.10. (Krummenacher, 1971). If the metering is to be done at the inlet to the device, then A is the outlet from a standing burette and the receiving ampoules B need not be calibrated. If the metering is to be done at the outlet, then A can be the outlet from a reservoir and the receiving ampoules B must be of the calibrated type described. The several phials attached to the manifold C are filled successively by moving the trough D along the manifold by means of the glass-enclosed iron core E attached to it and a magnet F.

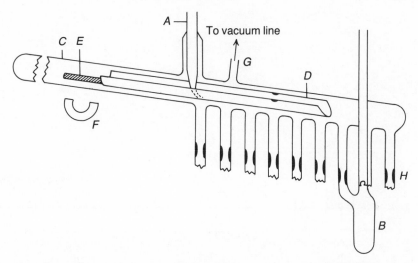

Fig. 3.10. Moveable trough for filling ampoules. *A* outlet from standing burette or reservoir, *B* phials, *C* manifold, *D* trough, *E* iron core coated with glass, *F* magnet, *G* connection to high vacuum system and *H* seal-off points.

3.1.5. Metering compounds by volume of vapour

A generally useful method of metering small quantities of compounds with 'reasonable' vapour pressures is to condense the vapour of the compound from a known volume at a known temperature. If the vapour of a liquid or a solid in a reservoir *A* can expand into an evacuated bulb *B* of volume *V* (see Fig. 3.11), then the number *n* of moles in *V* is given to an adequate degree of accuracy by the Ideal Gas Law $pV = nRT$ where *p* is the vapour pressure of the compound at a temperature *T* which must be lower than the ambient temperature T_1 of the vapour (Biddulph and Plesch, 1959).

For all polar compounds, which effectively means all compounds except hydrocarbons, the glass surfaces involved in measurements of this type must be made hydrophobic. This can be done by rinsing them with a solution of hexadecyl trimethyl ammonium bromide (Norrish and Russell, 1947) or, more conveniently, by filling the apparatus with the vapour of trimethyl chlorosilane, leaving it for 10–20 min, and then pumping off the excess of silane and the hydrogen chloride formed by reaction of the silane with the water on the glass surface. Both treatments yield what is effectively a hydrocarbon coating on the glass which does not adsorb polar compounds as strongly as glass does.

The measurements can be done in two ways, according to circumstances. If the quantity of compound to be measured out is between *ca.* 0.05 and 0.5 g, the contents of the bulb *B* of (approximately) known volume can be

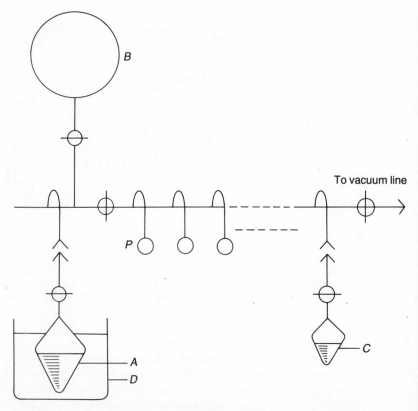

Fig. 3.11. Schematic drawing of assembly for dosing water into phials by volume of vapour. *A* reservoir containing distilled, deaerated water, *B* dosing bulb of known volume, *P* breakable phials, *C* container holding Me_3SiCl, *D* thermostat bath.

condensed into a phial *P* and weighed. This procedure can also serve to calibrate the bulb accurately. For such calibrations it is advantageous to use compounds of high molar mass and high vapour pressure, e.g. CCl_4.

For metering very small quantities, the volume of the bulb must be known accurately. The amount of material in it can be controlled very closely by changing the temperature of the compound in the reservoir, provided that its vapour pressure curve is known. By using the vapour of ice, it was possible to measure out *ca.* 10^{-5} mole of water (*ca.* 2×10^{-4} g), (Biddulph, Plesch and Rutherford, 1965).

It is evident that the vapour method cannot be used with compounds which dissociate appreciably at ambient temperatures, e.g. PCl_5 and $SbCl_5$.

If fixed quantities of a compound are required frequently, as when the rate of a reaction is studied as a function of a concentration, it is convenient to have a set of suitably sized dosing bulbs of, say, 5, 10, 20 ml, and to keep the compound in A at a fixed temperature, e.g. in ice.

3.1.6. Dilatometers (Plesch, 1986)

3.1.6.1. General principles Dilatometers are devices for measuring changes in the density of solids or liquids. Since virtually all physical and chemical changes are accompanied by a change of density, dilatometry can be used for many different purposes (Fries, Lewis and Weissberger, 1963). Its principal uses can be divided into measurements of density changes due to variation in physical conditions such as temperature or pressure, and of changes due to chemical reaction. As an example, we shall describe the use of dilatometers to follow the progress of polymerisations or depolymerisations, but the devices described below will be found useful for a wide range of liquid phase reactions.

A dilatometer consists essentially of a glass tube (the 'body'), containing the reaction mixture, to which a straight capillary tube is attached (usually, but not always, vertical). The movement of the meniscus of the reaction mixture in the capillary amplifies changes in the volume of the reaction mixture and is measured, as a function of time, with a cathetometer or by some automatic device. Dilatometry has been used for a very long time because it is simple, cheap, convenient and can be made very accurate.

The most important practical requirements which a dilatometer must fulfil are that it must be easy to fill and that the body must be so shaped as to make heat-exchange with the thermostat fluid adequately rapid. In other words, the enthalpy of any reaction taking place must be removed so efficiently that thermal expansion of the reaction mixture is negligibly small. It is this need to preserve strictly constant temperature (usually better than ± 0.01 K) which makes dilatometry unsuitable for reactions with half-lives of less than *ca.* 10 min unless the surface-to-volume ratio of the dilatometer body is exceptionally large. Dilatometers are very sensitive thermometers and particular attention must always be paid to thermostatting. It is generally useful for the body of the dilatometer to have a large surface-to-volume ratio and to construct the body from thin-walled glass tubing. There are many different designs for the body of dilatometers aimed at increasing the surface-to-volume ratio, such as the mushroom dilatometer (Burnett, 1954), and the tubular design (Pepper, 1949).

If the dilatometer body has one access only, namely through the capillary, the filling process is laborious and slow under atmospheric pressure because of air-locks in the capillary, but it can be swift and easy when done under vacuum. Emptying the body can prove difficult after a polymerisation reaction since the reaction mixture becomes very viscous. Therefore it is common practice to cut open the body after a reaction and to repair it for the next. The best way to break open the body of a dilatometer is to score a cross in the base of the body and to tap the middle of the cross with a wooden stick. Repairing the body of a dilatometer is considerably easier than sealing a new body onto a fine capillary and with the method described here, the expensive precision capillaries which are required for the most accurate work can be reused almost indefinitely. For some purposes it is useful to fit the capillary to the body by a cone joint, but in order to avoid the trapping of vapour bubbles the cone (pointing upwards), not the socket, must be fitted to the body. With such a device the cone should be sealed with grease only near its base so that the reaction mixture does not become contaminated by grease.

When the rate of a polymerisation is followed dilatometrically, i.e. by the change in density of the reaction mixture, one makes use of the fact that most monomers are less dense than the corresponding polymers so that the polymerising mixture contracts as the reaction proceeds. Provided that for any one initial monomer concentration m_0 there is a unique relation between the degree of conversion $Y = (m_0 - m)/m_0$ and the density ρ of the solution, (where m is the monomer concentration at any time t), and provided that the density of the polymer is independent of the degree of polymerisation, DP, of the polymer (which is a reasonable approximation above about $DP = 5$), then

$$d\rho/dt = f(dY/dt) \tag{3.2}$$

and if the relation of ρ to Y is rectilinear, then

$$d\rho/dt = k'dY/dt = -k''dm/dt \tag{3.3}$$

Since the change in the height of the liquid level in the capillary of the dilatometer is proportional to the change of volume of the reaction mixture, which, in turn, is proportional to the change in its density,

$$-dm/dt = -Adh/dt \tag{3.4}$$

where A is an apparatus constant. If the polymerisation is internally of first order, so that

$$-dm/dt = k_1 m \tag{3.5}$$

it follows that $-dh/dt = Ak_1 m$. If the initial value of h is h_0 and its final value h_f, then

$$\ln[(h - h_f)/(h_0 - h_f)] = -k_1 t \tag{3.6}$$

so that for such polymerisations A need not be determined, and any conventional treatment of the (h, t) readings will yield the rate constant k_1.

For any polymerisations which are not internally of first order, the apparatus constant A must be determined. This can be done as follows: Let the volume of solution in the dilatometer be $V_B + \pi r^2 h$ (where V_B is the volume of the body of the dilatometer, r the i.d. of the capillary and h the height of the meniscus in the capillary), and let the dependence of the volume

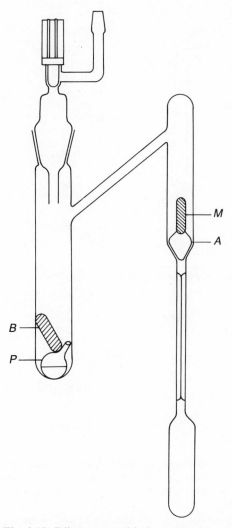

Fig. 3.12. Dilatometer with dry valve A containing iron core M; P is a phial of reagent, B a phial breaker.

on the extent of conversion Y be linear and such that when $Y = 0$, $h = h_0$ and when $Y = 1$, $h = h_r$. Then, at any stage of the reaction, the relation between the volume of solution and the monomer concentration is given by

$$Y = (h_0 - h)/(h_0 - h_r) \tag{3.7}$$

Therefore
$$-dm/dt = [m_0/(h_0 - h_r)]\,dh/dt \tag{3.8}$$

and
$$A = m_0/(h_0 - h_r) \tag{3.9}$$

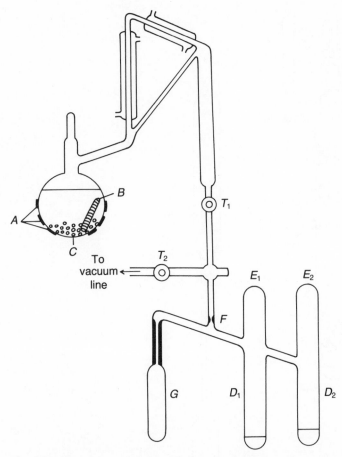

Fig. 3.13. The parts of this diagram above and below T_2 are not drawn to the same scale. The solvent supply system above tap T_1 is the same as that shown in Fig. 2.14 and explained there. A are the coils of a heating tape, B is a stirring bar to agitate the CaH_2, C.

When used for a polymerisation, the mixing chambers D_1 and D_2 are charged, respectively, with monomer and initiator (both crystalline) and the chambers are sealed at E_1 and E_2. The assembly is evacuated and part of the solvent is run from T_1 into D_1, part into D_2. These are then frozen and the assembly is fused off at F. After thermostatting, the solutions in D_1 and D_2 are mixed to start the polymerisation and the mixture poured rapidly into the dilatometer body G (Hamid, Novakowska and Plesch, 1970).

From a consideration of the above argument it is clear that A must be determined for each dilatometer and each m_0. This is generally simpler and more accurate than calculating A from the dimensions of the dilatometer and the density of the solutions concerned.

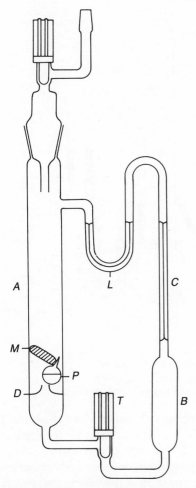

Fig. 3.14. Tap dilatometer with loop, L, for stopping distillation into the capillary, C, of the dilatometer, B; A mixing chamber, M phial breaker, P phial, D trap to retain phial fragments, T PTFE tap.

3.1.6.2. A variety of models   If the most rigorous technique is not required, for example, if a monomer is to be polymerised at, say, 80 °C by a radical initiator, then solvent and monomer are run into the mixing chamber, the catalyst is added and left to dissolve, the assembly is then attached to a vacuum line to allow the reaction mixture to be degassed by the conventional freeze–pump–thaw process and to facilitate the filling of the body and the capillary. When this has been done, the dilatometer is thermostatted and the height of the meniscus in the capillary is monitored by means of a catheto-meter. The simple dilatometer adequate for this can be modified for more

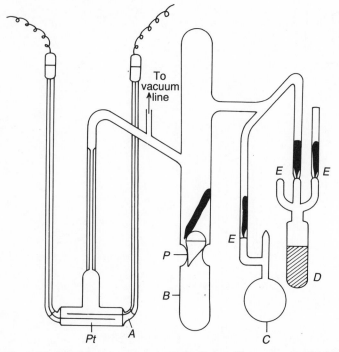

Fig. 3.15. An elaborate reactor attached to a dilatometer *A* which is fitted with electrodes (*Pt*) for conductance measurements. *B* mixing chamber, *P* phial of styrene, *C* bulb containing HI gas, *D* solution of iodine, *E* break-seals.

rigorous work by introducing the initiator in a sealed glass phial and the solvent and monomer by distillation from a vacuum line after the apparatus has been evacuated (Fig. 3.12).

If one is using a solvent of relatively high vapour pressure the solvent will tend to distil from the residual reaction mixture in the mixing chamber into the capillary because of the difference in vapour pressure between surfaces of different curvature. This can be prevented by incorporating a dry-joint (see p. 47) between the top of the capillary as shown in Fig. 3.12, or by incorporating a small length of capillary with the same i.d. as that used for the dilatometer, as shown in Fig. 3.14.

A variant with two mixing chambers is shown in Fig. 3.13, which also shows one type of solvent delivery system (Hamid, Nowakowska and Plesch, 1970). The advantage of having two mixing chambers is that it allows a solution of the initiator to be made which can then be mixed rapidly with the monomer solution to start the reaction. This is very much more efficient than mixing either of the components undiluted with a solution of the other.

Another type of dilatometer, designed to allow rapid filling of the body, is shown in Fig. 3.14 (KPG). With this type of dilatometer the reaction

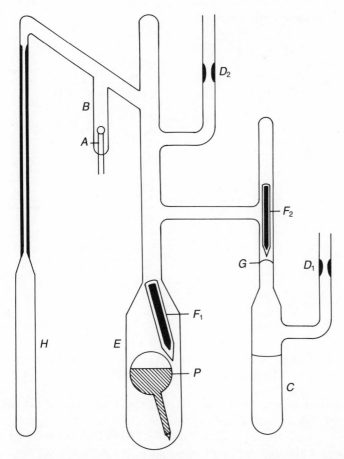

Fig. 3.16. Dilatometer with electrodes. The Pt electrodes A are fitted to an appendix B. The apparatus was used for the polymerisation of styrene (distilled into C, followed by solvent) by trifluoroacetic acid (in phial P). After being charged with styrene, D_1 is sealed off, solvent is distilled into E, then D_2 is sealed off, P is broken with F_1, G is ruptured with F_2, and the reaction started by mixing the contents of C and E; the dilatometer body H is filled rapidly, some of the solution being left in B.

mixture is transferred into the body of the dilatometer through a PTFE tap. Experience has shown that this tap does not cause any difficulties.

It is often useful to follow the electrical conductance of the reaction mixture (Giusti, 1965; Obrecht and Plesch, 1981), and for this purpose, a pair of Pt electrodes can be sealed into the body (Fig. 3.15 (Giusti, 1965) and Fig. 3.17 (KPG)), or into an appendix on the bridge-tube (Obrecht and Plesch, 1981) (see Fig. 3.16).

The elaborate reactor in Fig. 3.15 was designed for studying the

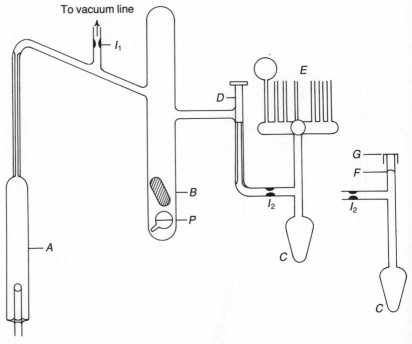

Fig. 3.17. Dilatometer with two alternative devices for sampling the reaction mixture during a reaction. Detailed explanation in the text.

polymerisation of styrene by I_2 and HI; that of Fig. 3.16 for the same monomer with trifluoroacetic acid as initiator.

If UV-vis. spectra are to be taken, e.g. at the end of the reaction, without exposing the reaction mixture to the atmosphere, an extra arm terminating in a quartz cell can also be added to the assembly. Many such devices can be found in the literature, and three in this book (Fig. 3.25, 3.26, and 5.8) which can be combined with a dilatometer.

In many investigations it is useful, or even essential, to examine the products formed at different stages of the reaction. In Fig. 3.17 a dilatometer is shown whose body A is fitted with electrodes and which has a sampling vessel C attached to the mixing chamber B. This sampling device has also been used without a dilatometer (Jones and Plesch, 1979). After it has been charged with reagents and solvent, the whole rig is separated from the vacuum line by sealing off at I_1. When the reaction has been started by rupture of the phial P of initiator, the reservoir C of the tipping device is filled, the PTFE tap D is closed, the contents of C frozen and the tipping device sealed off at I_2. The phials E are then filled and thermostatted and the dilatometer A is filled and thermostatted. The phials E can be sealed off at any convenient time and analysed at chosen times. An alternative

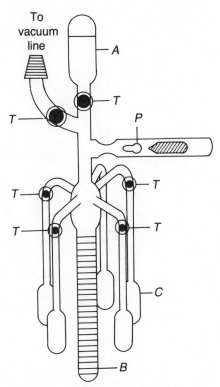

Fig. 3.18. Arrangement for filling several dilatometers successively with the same stock solution of monomer. This is suitable only for very slow reactions. Details of operation are in the text. All the taps T are PTFE taps.

sampling device is shown on the right of Fig. 3.17. After it has been fused off from the dilatometer, its contents are brought to atmospheric pressure by punching a thick syringe needle carrying a stream of inert gas through the rubber septum G and the break-seal F. Samples of the reaction mixture in C can then be withdrawn with the syringe.

An apparatus for filling several dilatometers in one operation is shown in Fig. 3.18 (Brzezinska, Matyjaszewski and Penczek, 1978). In the original work a solution of monomer was prepared in the chamber A separated from the dilatometers and hanging burette B by a PTFE tap. The initiator phial P was then broken and the initiator washed into the burette with a given portion of the monomer solution from A. After tipping a portion of the reaction mixture into one of the dilatometers, C, more monomer solution could be added to the mixture remaining in the burette and the second dilatometer filled. In this way up to six dilatometers could be filled with solutions having identical monomer concentrations but different initiator concentrations. Obviously, it is not difficult to devise variations of this

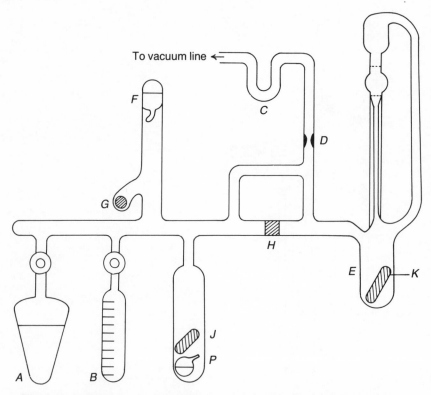

Fig. 3.19. Vacuum viscometer design. *A* solvent reservoir, *B* hanging burette, *C* site for freeze seal, *F* killing agent in break-seal phial, *G* stainless steel ball bearing in appendix, *H* coarse frit to retain glass fragments, *J* phial breaker, *K* stirrer, *P* initiator phial.

technique, which can increase the number of experiments which can be done in a given time.

Finally, although it will not be described in detail here, an ingenious instrument worth noting is a viscometer-dilatometer (viscodilatometer) which, although not designed to work under vacuum, can be modified for this purpose (Gordon and Grieveson, 1955).

3.1.7. Viscometers

It is occasionally useful to monitor the viscosity of a reaction mixture as the reaction progresses, especially if the reaction concerned is a polymerisation or depolymerisation. A vacuum viscometer was designed to function without any taps and with an arrangement for mixing and transfer of materials which would prevent fragments from a broken glass phial entering the capillary of the viscometer (Krummenacher, 1971). A new design derived from it by the author is shown in Fig. 3.19. Both designs incorporate an Ubbelohde suspended-level viscometer. The derived design, being equipped with PTFE taps, is more flexible but less rigorous than the original, which illustrates again the Second Law of Experimentation. Both devices are suitable only for very slow reactions.

The design in Fig. 3.19 is intended to be used thus: Solvent is distilled from the vacuum line into A, and a measured amount of monomer is distilled into B. Then a freeze seal is made with solvent at C and the viscometer is sealed off at D. All the monomer is distilled from B to E, solvent is distilled from A to B and a measured amount of it first into E to dissolve the monomer and more onto the phial P of initiator which is then broken. If the initiator is volatile, this is not necessary and it can be distilled directly into E. If it is not volatile, a solution of it is poured into E through the frit H which retains the glass fragments. After good mixing in E, the whole rig is thermostatted and the viscosity measurements are started.

The effect of dilution on reaction rate can be studied by distilling more solvent into E. The reaction can be stopped by breaking the seal of F with G and cooling E to distil the killing agent into the reaction mixture.

3.1.8. UV-vis. spectroscopic measurements

The range of qualitative and quantitative information available from the spectrum of a solution has led to a variety of designs of apparatus for measuring such spectra without exposing the material to the atmosphere. One of the commonest reasons for recording UV–vis. spectra of solutions under vacuum is to determine the concentration of an ion. In the context of anionic polymerisations it may be the naphthalide ion or the polystyryl carbanion, in the context of cationic polymerisations it may be a triarylmethyl carbenium ion, and consequently many variants of the required basic device have been published. In connection with cationic polymerisations, the quantity of acid, e.g. $HClO_4$, or of an aroyl salt, e.g. $ArCO^+SbF_6^-$, in a phial can be determined easily by letting it generate the trityl ion from Ph_3COH, and determining this by means of its well-known extinction coefficient. ($\lambda_{max} = 413$, 433 nm, $\epsilon = 3.8 \times 10^4$ l mol^{-1} cm^{-1} (Kalfoglou and Szwarc, 1968). For measurements to determine the concentration or the quality of an ionogenic compound (e.g. a strong acid) or of an ionic compound, the basic device must comprise the following features: A facility for introducing a sealed phial, an arrangement for dissolving or diluting the contents of the phial, a facility for introducing a second reagent in a phial or otherwise, and a controllable supply of solvent to make and/or dilute the required solution and, finally, a suitable optical cell, and, of course, the overriding requirement is that the whole rig is shaped so that the cell can be fitted into a spectrometer. A very early and elaborate apparatus that meets these requirements is shown in Fig. 5.8. Rather simpler designs are feasible, especially if PTFE taps are permissible, and two devices comprising optical cells are described in Section 3.2.2.

Since optical cells for UV work must be of quartz, and since this cannot be sealed to borosilicate glass, a tube consisting of a series of zones of progressively changing composition, called a 'graded-seal', must be interposed. Cells with such graded-seals are available commercially, but they are expensive and fragile. The graded-seals should not be heated unevenly and the optical cells are best not heated at all. When joining such cells to an apparatus always allow at least *ca.* 2 cm between the graded-seal and the join, and before starting work wrap the cell and the graded-seal with ceramic string to protect it from a chance encounter with the flame of a blow torch. Because of the expense involved, especially careful planning of the apparatus design with respect to the protection of these fragile appendages is strongly advised.

For measurements at temperatures other than ambient, cells with double walls, which can be thermostatted, are also available commercially. If measurements are required at temperatures between *ca.* −5 °C and room temperature, the sample compartment of the spectrometer can be flushed with dry air or nitrogen to reduce condensation on the cell windows. Below *ca.* −5 °C the windows can be covered with a thin polythene film, but measurements below −25 °C are very troublesome. The problems associated with low temperature spectroscopic measurements were solved by enclosing the cell in an air-tight box fitted with glass windows (Dadley and Evans, 1967). The box was so designed that it fitted into the spectrophotometer and the air inside the box was dried with phosphoric oxide which, it is claimed, stopped condensation even at temperatures as low as −60 °C; glass windows could be used because only absorptions above 380 nm were of interest.

Recent developments in fibre optics have made it possible to measure the spectrum in any kind of vessel, including any which form part of a vacuum apparatus, so that the problem of fitting a cell which is part of a vacuum rig into a spectrophotometer is now effectively obsolete. However, for single or occasional measurements the devices discussed here are simpler and very much cheaper than fibre-optic gadgetry.

3.1.9. NMR measurements

Measuring the NMR spectra of samples prepared and sealed under vacuum presents no special problems. NMR tubes available commercially are usually made from borosilicate glass and can therefore be fused to normal apparatus for filling. The seal-off constriction above the NMR tube should be made carefully to ensure that sealing-off can be done easily and leaves a thickened end which is as centralised as possible. If the seal-off is asymmetric, the NMR tube will precess when spun during a measurement. Precession can also be induced by larger glass fragments being washed into the sample tubes during filling and the constriction should therefore leave an opening only *ca.* 1 mm wide to act as a coarse filter. In order to allow the expensive NMR tubes to be reused, a length of narrow tubing (o.d. *ca.* 4.0 mm, wall thickness *ca.*

1 mm) can be sealed to the NMR tube, but this join is best prepared professionally on a lathe to ensure that the extension and the NMR tube are strictly co-axial. An alternative is to make the NMR tubes from standard glass tubing. This is a simple process and, with practice, the spectra obtained will not be significantly different from those obtained with commercial tubes.

To make NMR tubes, one first chooses a length of thin-walled glass tubing with an o.d. slightly greater than that required (i.e. *ca.* 5 or 10 mm). The tube is then cut into lengths of *ca.* 20 cm and one end of each piece is sealed and rounded. The tubes are then soaked in a 30% solution of HF to etch them until they fit exactly into the NMR spinner mechanism. (*Attention*: HF is extremely corrosive and spillages must be treated immediately.) In order to check whether the tubes have the correct diameter, they are removed from the HF bath and rinsed with a 10% $NaHCO_3$ solution and then with distilled water. Only after a thorough wash should the diameter be measured with a vernier caliper. When the diameter is correct, the tube must be cleaned thoroughly and dried, and finally tested in the NMR spinner. The time required to etch the tubes to the correct diameter depends on how near to the desired diameter the original tubes are, but suitable tubes can usually be prepared within several hours.

It is often desirable to have tetramethyl silane as the reference compound for NMR measurements, and also many modern spectrometers require a deuterated solvent for a locking signal. Adding tetramethyl silane to the reaction mixture is not always desirable, especially if NMR is not the only means of analysis being used, and deuterated solvents are very expensive. Both these problems can be circumvented by incorporating an insert in the NMR tube before attaching it to the vacuum apparatus. A normal boiling-point tube can be used as the insert for 5 mm NMR tubes and the 5 mm NMR tubes themselves make excellent inserts for 10 mm NMR sample tubes. These inserts are filled with the desired reference material and their open ends sealed. The insert is then fitted into the sample tube and held central by two PTFE washers; these are available commercially, but they can be fashioned easily from the PTFE of an unusable vacuum tap. It is important to check that inserts made as described do function as such before sealing them into a vacuum apparatus.

The preparation of samples for NMR measurements totally *in vacuo* should present no difficulties for a chemist used to operating with h.v.t., and therefore the necessary apparatus will not be described here.

3.1.10. Conductance and other electrical measurements

3.1.10.1. Electrical conductance   One of the easiest measurements to make, but one which is often ignored, is the electrical conductance of a solution. In contrast to UV–vis. spectroscopy, which can only be used to monitor species having a suitable chromophore, all species carrying a net charge (single and triple ions, not ion-pairs etc.) can be monitored via the conductance of their solutions. The conductance G (reciprocal of resistance) of a solution is measured by applying an alternating potential to two electrodes, usually of Pt, immersed in the solution and measuring the resistance with some kind of bridge and detector. G depends on, amongst other factors, the geometry of the electrodes, characterised by a so-called cell constant, A, so that the conductivity which is characteristic of the solution itself, is given by

$$\kappa = G/A \qquad (3.10)$$

A also depends on the depth of immersion of the electrodes up to a certain value and the electrodes should always be at such a depth that A is independent of it. In most modern instruments either the electrodes are constructed to have $A = 1$ cm or there is a facility for compensating for values of A other than 1 cm. For accurate work the cell constant is usually measured for each set of electrodes by a standard method (Lind, Zwolenik and Fuoss, 1959).

The conductivity of an ionic solution is given by

$$\kappa = \sum \lambda_i c_i, \qquad (3.11)$$

where the summation includes all charged species in the solution and λ_i and c_i are the ionic conductivity and the concentration of the ith species. The λ is determined by the charge, size and shape of the charged species, which may be an ion or an aggregate of ions with a net charge, and λ also depends on the viscosity and the permittivity of the solvent, and through these quantities on temperature and pressure; is also depends on the ionic strength of the solution and on the frequency of the AC current used for the measurements.

Conductivity is usually measured for one of two purposes: It can be used as an indicator of changes in the ionic population of a solution during a reaction and, if done very precisely, with many precautions, and often at different frequencies, it can provide a fundamental understanding of the ionic solution under test. Here we are concerned mainly with the first use, and so the conductivity changes which concern us have a status similar to that of temperature or colour changes as indicators of what is going on in a solution.

A simple conductivity probe is shown in Fig. 3.20 (KPG). This probe can be used for both rough and precise measurements, but when used for precise measurements of low conductances it may give rise to capacitance effects. To avoid large capacitances, the leads to the conductivity cell should be situated as far apart as possible (see, e.g. Fig. 3.15), and they must be screened.

The reason for the graded-seal in Fig. 3.20 is that thin Pt wire forms a vacuum-tight seal with soda glass but not with standard borosilicate glass. The Pt/soda glass seal is, however, fragile due to the difference in the thermal expansion coefficients of soda glass and Pt. Thus, for the glass spacers lead

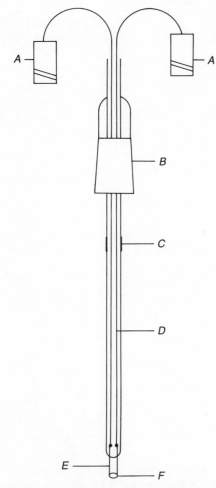

Fig. 3.20. Common conductivity probe for vacuum work. *A* couplings to
shielded cables from conductivity meter; *B* a B.19 or B.24 socket fitting onto
cone on observation/reaction vessel; *C* graded-seal soda glass to borosilicate
glass; *D* shielded leads spot-welded to thick Pt wires *E* sealed through the soda
glass probe and held together by the lead glass bead *F*.

glass, with a thermal expansion coefficient closer to that of Pt, is used. This
simple conductivity probe is useful for fitting into small reaction vessels by
a cone and socket joint, as shown, or it can be fused into a port on a reaction
vessel. A variation on this simple design allows the length of the probe to be
adjusted, and is very suitable for less exacting work where the ground-glass
joints are tolerable (Plesch, 1973) (Fig. 3.21). To avoid the necessity of a
graded-seal, the Pt wire of the conductivity cell can be spot-welded to
tungsten rods which give a good seal with borosilicate glass; such a design is

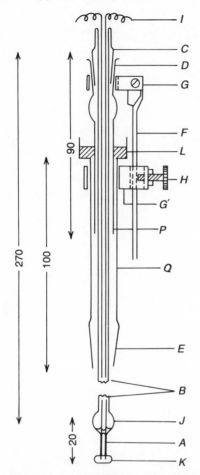

Fig. 3.21. The PLADE-C adjustable depth electrode assembly. The gap between the plunger P and the barrel Q of the syringe is exaggerated for clarity; the rubber pads protecting the glass from the metal collars G and G' are not shown. The dimensions (in millimetres) shown in the figure are approximate and are not critical. Pt wire electrodes A are sealed through a soda glass tube B which ends in a cone C (preferably B.10). This cone C fits into a socket D fused to the top of the plunger P of a standard hypodermic syringe (such as a 'Chance' 4 cm³ syringe) of Pyrex glass, and it can be secured there by a suitable sealant. The lower end of the syringe barrel Q is fused to a B.14 cone E which fits into a socket on the reaction vessel. In order to prevent the plunger from being sucked in or blown out when the electrode is used at a pressure other than atmospheric, a clamping device is used. This consists of a strip of metal F, fixed to the plunger by a metal collar G, which can be clamped by a screw-clamp H fixed to the top of the barrel by another metal collar G'. Electrical contact is made through two copper leads I spot-welded at J to the Pt wires A, whose distance apart is fixed by the lead glass bead K. For applications where lubricants cannot be tolerated, the cup L is filled with mercury, which then provides a conventional mercury seal for the plunger. The spilling of mercury can be prevented by a wad of cotton wool.

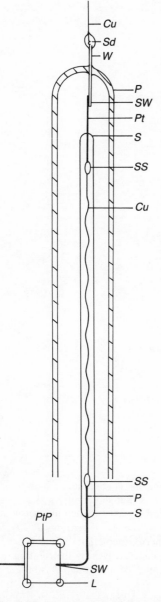

Fig. 3.22. Arrangement of the leads inside the conductivity cell shown in Fig. 3.23. *Cu* copper wire, *Sd* soldered joint, *W* tungsten wire, *SW* spot-weld, *Pt* platinum wire, *S* soda glass sleeve, *SS* silver-soldered joint, *P* borosilicate glass arm fused to the cell, *PtP* platinum plate electrodes held together with lead glass beads *L*.

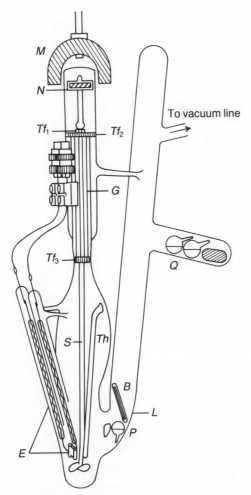

Fig. 3.23. Conductivity cell with phial magazine Q containing phials and magnetic pusher. The electrode assembly E is that shown in Fig. 3.22. S is a stirrer shaft with a propeller at one end and a glass-enclosed magnet N at the other. The stirrer shaft is held in position by the PTFE bearings Tf_1, Tf_2, and Tf_3 and the glass tube spacer G; Th is a thermocouple pocket and B a magnetic breaker for the phial P. The propeller, driven by the rotating magnet M, pumps the cell contents around the loop L, so that when P is broken there is very fast mixing.

shown in Fig. 3.22 (Grattan and Plesch, 1977). In the same publication a complete apparatus for measuring conductivities *in vacuo* is described which incorporates several generally useful features (Fig. 3.23). The cell was stirred from above by a rotating permanent magnet through a magnetic clutch. This arrangement can be made considerably more efficient for apparatus of irregular shape than stirring from below with a magnetic bar inside the apparatus (see Section 3.3). The pear-shaped reaction vessel is an important feature because it allows the electrodes to be covered adequately when the vessel contains only *ca.* 50 ml, but allows volumes of up to 500 ml to be used, so that a ten-fold dilution is possible. The phial magazine and the phial breaking device are also of general applicability. When kinetic measurements are being made, it is important that the phial is allowed to equilibrate thermally with the solution before being broken, and that the phial contents are dispersed rapidly throughout the mixture.

Devices used for conductivity measurements in the context of anionic polymerisations have been described by many authors (e.g. Worsfold and Bywater, 1960; Bhattacharyya, Lee, Smid and Szwarc, 1965; Fig. 5.8). The later paper also contains many other useful vacuum devices, and many derived designs have been published since.

For accurate measurements of conductivity under 'normal' conditions it is usual to use platinised Pt electrodes because such electrodes are less affected by polarisation, but this type of electrode is not only difficult to reproduce when it is incorporated into a complex vacuum apparatus, it is difficult to clean, not very robust, and can produce degassing problems. Therefore, they are not recommended for conductivity measurements *in vacuo.*

A note on units   In the older literature, and even now in some books, the Equation (3.11) is written as

$$10^3\kappa = \sum \lambda_i c_i \tag{3.12}$$

because c was given in mol l^{-1}, and λ in S cm^2 mol^{-1}, so that κ has the units S cm^{-1}.

From a long experience this author advises strongly that anyone doing conductivity measurements should adhere strictly to SI units and use Equation (3.11). Then, with A in m, c in mol m^{-3} (numerically the same as millimoles l^{-1}), κ is in S m^{-1}, and λ in S m^2 mol^{-1}.

3.1.10.2. Other electroanalytical methods   The use of h.v.t. in conjunction with electroanalytical techniques of the potentiometry-polarography type has been described in detail (Kesztelyi, 1984), so that it need not be discussed here. That author, however, ignores a very useful cell for electrosynthesis under vacuum (Schmulbach and Oommen, 1973) and the electrochemical techniques developed by Szwarc and his co-workers and others in the context of anionic polymerisation, which we have mentioned above.

3.2. Combined measuring devices

3.2.1. General considerations
The economically-minded chemist attempts to glean the maximum of information with the minimum expenditure of work, time and money, and for this reason much effort has gone into devising apparatus by means of

which more than one kind of measurement can be made on each reaction mixture, preferably simultaneously. The most frequent type of measurement, and usually the most important one, is of conversion or yield as a function of time, i.e. a kinetic measurement. Many of the devices described in the earlier sections of this chapter are devised, or can be used, for making kinetic measurements by a single technique, and the aim of this section is to describe devices whereby several types of measurement can be made on the same reaction mixture. Very good examples of such devices are the dilatometers fitted with electrodes, which have been described in Section 3.1.6. rather than in this section for the sake of economy and continuity.

In kinetic measurements the temperature usually enters the picture as a quantity to be kept as constant as possible, and the usual reactors are therefore classed as isothermal. However, it is very difficult to keep very fast reactions isothermal because of the difficulty of removing (usually) or supplying (very rarely) the enthalpy of reaction. But, with sufficient ingenuity, one can often put to good use a difficulty which one cannot avoid. As far as we know, the rate of temperature rise due to the enthalpy of reaction under adiabatic conditions was first used by various members of M. Polanyi's laboratory in Manchester in the 1940s. They recognised that if a reaction is too fast for the reaction mixture to be kept at constant temperature, then if the reation is done adiabatically, e.g. in a Dewar vessel, the rate of temperature rise gives directly the rate of reaction (Evans, et al., 1946). This is the basis of kinetic calorimetry. An exact mathematical treatment of adiabatic kinetics was first given by Gordon (1948).

Some of the most generally useful isothermal and adiabatic reactors and a few other devices, all usable under high vacuum conditions, will be described in the following sections.

3.2.2. Isothermal reactors
The simplest type of high vacuum isothermal multi-functional reactor is any vessel immersed in a thermostat, or fitted with a jacket through which fluid from a thermostat is pumped, which can be charged on a vacuum line, and on whose contents two or more types of measurement can be made. Most reactors of this category are little more than round-bottomed (or other robust) flasks with several necks or ports, which carry various probes and feed lines (vacuum, solvent, etc.). There is, however, one device which, although its originator used it for preparative rather than kinetic studies, became one of the inspirations for the Pask–Plesch adiabatic calorimeter (Pask and Plesch, 1981) and which, because of its elegance and potential for development, may appeal to others; this is the Krummenacher 'Universal-reaktor', and because it has been published only in a thesis we will describe it in detail (Fig. 3.24). It was designed for the polymerisation of 4-phenyl-dioxolan by perchloric acid (Krummenacher, 1971a).

The body of the reactor is a double-walled vessel A (o.d. 5 cm) with two

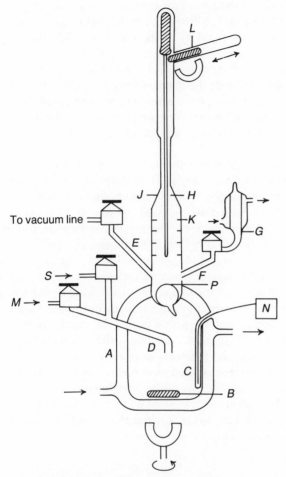

Fig. 3.24. The Krummenacher reactor. Most of the parts are explained in the text. To start a reaction, the retainer L is withdrawn magnetically so that H drops and breaks P which contained the initiator. The thermocouple C and the conductivity probe (not shown) were linked to the same recorder N. All the taps were all-metal BiPl valves (see Sections 2.2.4.3 and 2.4).

access ports to the space between the walls; it can therefore be used as an isothermal reactor if fluid from a thermostat is pumped through this space, or as an adiabatic reactor if this space is evacuated. The contents of the chamber, diameter 3 cm, are stirred by a bar magnet B driven by a rotating magnet. The temperature was monitored by a temperature sensor inserted into a pocket C. This arrangement is a weakness which detracts from the utility of this device for adiabatic kinetic measurements, because for these the bare temperature sensor (usually a Pt resistor) should be immersed in the reaction mixture in order to minimise heat-transfer delays. The conductivity electrodes (not shown) were of thin Pt wire, their ends held together by a lead glass bead, insulated by glass capillary tubing and spot-welded to tungsten rods sealed through the borosilicate glass. Two disadvantages of this construction are that the 'cell constant' of this probe varies with the depth of solution in the vessel, and that the reaction mixture is difficult to remove from between the wires and their glass sleeves. The weaknesses of this device are mentioned deliberately to prevent others making mistakes which, whilst difficult to avoid in 1970, in a laboratory unfamiliar with h.v.t., are easy to avoid now.

Access to the reactor is through three ducts, one of which, D, is for monomer supply (M) and solvent supply (S) through the same duct, one, E, is for vacuum and one, F, for the 'killing' mixture (in G) to neutralise the initiator and the growing chain-ends. Actually, all four services could have gone into the reactor by a single duct. The most prominent feature is the central phial-breaker H and phial-holder J. At the end of a polymerisation the phial-breaker assembly was cracked off at one of the prepared notches K by hot-spotting, the reactor was emptied by means of a syringe, washed, charged with a phial P of initiator solution and the breaker device was then fused on again. Evidently, the opening and remaking of this joint are tricky, and with the experience gained since this device was made in 1970 it is possible now to get just as rigorous conditions with a cone-joint at this position to give greater ease and flexibility of operation.

Two different isothermal reactors designed, not for kinetic measurements, but for the observation of equilibria by means of simultaneous spectroscopic and conductivity measurements were devised by two groups (Holdcroft and Plesch, 1985; Pask and Nuyken, 1983). The purpose of both was to observe how a binary ionogenic equilibrium of the type

$$A + B \rightleftharpoons Q^+ + R^- \tag{3.13}$$

in a solvent is affected (1) by dilution at constant [A]/[B], (2) by changing [A]/[B], and (3) by changing the temperature. The requirements are that different proportions of a solution containing A and B and of pure solvent should be brought together in the observation vessel, or that different proportions of solutions of A and B should be mixed in the observation vessel. The two devices provide an interesting instance of two different solutions to the same problem. Actually, the device from Szwarc's laboratory shown in Fig. 5.8 can be used for the same purpose.

Pask and Nuyken's device (Fig. 3.25) consists of a mixing chamber attached to, and detachable from, a vacuum line carrying two burettes whose contents can be discharged into it. After the mixing chamber has been charged, the device is removed from the vacuum line and the reaction mixture is distributed to the conductivity cell and the UV cell through the PTFE taps. Both cells are jacketed and the jackets are perfused

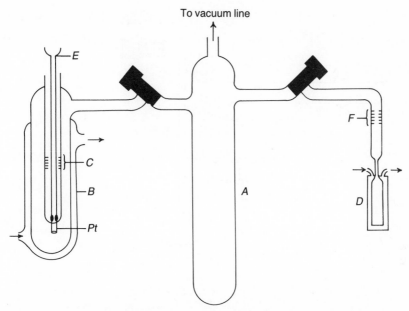

Fig. 3.25. The Pask–Nuyken device for measuring simultaneously the conductivity and the UV spectrum of a reaction mixture. *A* mixing chamber, *B* conductivity cell with jacket, *C* graded-seal borosilicate glass–soda glass, *D* jacketed quartz cell, *E* copper leads to platinum electrodes *Pt*, *F* graded-seal borosilicate glass–quartz.

by fluid from the same thermostat. The UV cell is inserted into the spectrophotometer and the UV and conductivity measurements are made. Then these can be repeated at different temperatures over the whole required range. After completion of this set of observations, the contents of both cells are transferred back to the mixing vessel, the device is reattached to the vacuum line, the composition of the mixture is changed by the addition of one reagent or of solvent from a burette, and the cycle of operations is repeated with the new mixture.

The Holdcroft–Plesch reactor (Holdcroft and Plesch, 1985), shown in Fig. 3.26, like the Pask–Nuyken device (Pask and Nuyken, 1983), needs to be detached from the vacuum line in order that the conductivity cell may be submerged in a thermostat or that the optical cell may be inserted into the thermostatted cell holder of the spectrophotometer. In contrast to the Pask–Nuyken cell, the two burettes containing either a solution of one of the reagents and pure solvent or solutions of both reagents, are an integral part of the assembly so that the composition of the reaction mixtures can be changed without having to reattach the assembly to the vacuum line. To see how the burettes were charged from the reservoirs on the vacuum line, the original paper should be consulted. It appears now that the most desirable

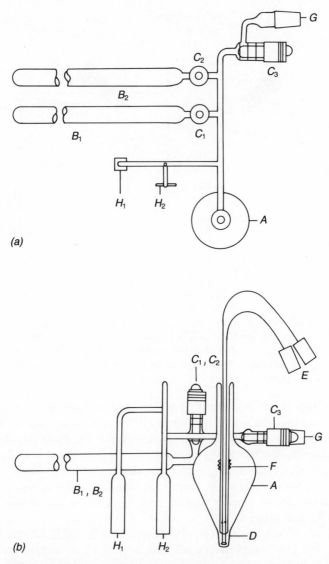

Fig. 3.26. The Holdcroft–Plesch observation vessel to find the conductivity and the UV spectrum of a solution; (a) plan, (b) elevation. A mixing vessel also serving as conductivity cell, B_1 and B_2 burettes, C_1, C_2, C_3 PTFE taps, D Pt wire electrodes fastened together by a lead-glass bead, E terminals from screened cables leading to D, F graded-seal soda glass–Pyrex, G cone for connection to vacuum and supply lines, H_1 1 cm and H_2 0.1 cm quartz cells.

device for combined conductivity and optical measurements would be a combination of both designs, incorporating the twin burettes of one and the jacketing of both observation vessels of the other; a further simplification could be made by inserting the conductivity electrodes into the optical cell, but keeping them clear of the light path (Bertoli and Plesch, 1968).

Note: The terms 'isothermal' and 'adiabatic' as applied to reactions (and reaction calorimeters) do not have the same meaning here as when used by thermochemists. The excellent introduction to reaction calorimetry by Skinner, Sturtevant and Sunner should be consulted (Skinner *et al.*, 1962).

3.2.3. Adiabatic reactors (reaction calorimeters)
The conversion of the Dewar vessel with a rubber bung carrying all the fittings, which had been used in Manchester in the 1940s as an adiabatic reaction calorimeter, to a versatile vacuum-tight device required two essential developments. With the original device it was difficult and awkward to do measurements below ambient temperature. The author overcame this difficulty by inventing the pseudo-Dewar vessel, i.e. one in which the Dewar space between the walls could be evacuated or filled with air at will (Plesch, Polanyi and Skinner, 1947). In this way the contents of the vessel can be cooled by immersing the charged vessel in a cooling bath whilst the Dewar space is full of air and then, when the contents have reached the desired• temperature, evacuating the Dewar space. The second requirement was to replace the rubber bung by a flanged head fitting on to a flange fused to the top of the pseudo-Dewar vessel, so that the vessel could be evacuated. The final stage of this development was the Biddulph–Plesch calorimeter (Biddulph and Plesch, 1959) with a fitting carrying conductivity electrodes (Gandini, Giusti, Plesch and Westermann, 1965). The construction and operation of this apparatus have been described by its originators and by at least two of the many others who used it (Blake and Eley, 1965; Bawn *et al.*, 1971); in fact, it became the most widely used adiabatic vacuum reaction calorimeter. Because descriptions of its construction and use are readily available, they will not be given here, especially as it has been largely superseded by the Pask–Plesch reaction calorimeter to be described below.

The weaknesses of the Biddulph–Plesch vacuum reaction calorimeter included the flanged joint, a rather large volume (not less than 70 ml), a rather brutal form of stirring which could set up destructive vibrations, and the rather delicate Pt wire temperature sensor. The first design of adiabatic calorimeter which was not an obvious descendant of the Biddulph–Plesch machine was made by Sigwalt and co-workers. (Cheradame, Vairon and Sigwalt, 1968; Favier, Fontanille and Sigwalt, 1974). Almost everything about it is different. Because it is free of joints it offered a new standard of experimental rigour and because thermal, spectroscopic and conductance measurements could be made simultaneously, its power was far ahead of other machines. However, since part of the device is of quartz and since each experiment requires the construction of almost a new machine, the chemist using this is entirely in the hands of the professional glass blower, the rate of experimentation is extremely low and the Second Law of Experimentation shows its power: the (almost) perfect experiment takes very, very long! Whereas with the Biddulph–Plesch and the Pask–Plesch calorimeters an average of four experiments per week can be kept up for several weeks, the rate of experimentation with the Sigwalt calorimeter is 3–4 experiments per month. As details of the construction and operation of this machine are given in readily available publications, and as it is unlikely to be adopted widely, we will not discuss it here.

The direct ancestor of Sigwalt's multi-purpose kinetic reactor is a preparative reactor originating in the same laboratory (Boileau, Champetier and Sigwalt, 1963). This is completely sealed and the solvent and all reagents are introduced through break-seals, and there are facilities for sampling the reaction mixture and for killing the samples separately. This too, is a glass blower's tour-de-force and unlikely to be adopted in many laboratories.

The most recent version of the Pask–Plesch adiabatic reactor (Pask and Plesch, 1981) is shown in Fig. 3.27. The original version had no ground joint in contact with the reactor space, being fused on to the vacuum line for each experiment; in a later version a sound ground joint was found to be equally effective and easier to use. Also, if required, the heater F, the conductivity probe G, and the temperature probe J can be inserted through ports with ground joints instead of being sealed into the vessel. This would make the calorimeter very much easier to construct, and there would be little, if any, reduction in rigour, especially if the cones were attached to the calorimeter and the fittings carried the sockets. In the original published version the Dewar space was evacuated once and for all and then sealed off at K. This implies necessarily that cooling must be done by refluxing the solvent by means of the cold finger C. Subsequent experience showed that this was not a useful change and that reverting to the author's pseudo-Dewar principle by having a tap at K (so that air can be admitted to the Dewar space and cooling effected by a cooling bath) is useful and indeed irreplaceable when working with involatile solvents, such as nitrobenzene.

The main advantages of the Pask–Plesch reactor are its small volume (30–50 cm^3 of solution), and the use of two commercial Pt sensors, one as a temperature probe and the other for the controlled heating which is necessary for the electrical calibration before and after each experiment. Both are Johnson-Mathey 100 Ω non-inductively wound flat elements with the Pt wire covered by a very thin ceramic layer. This is essential because the glass phial fragments being churned around by the stirrer during the operation often damage unprotected Pt wire. In the original publication a different type of heater was described which was later abandoned.

The initiator, and in many experiments a stopping (killing) reagent, was

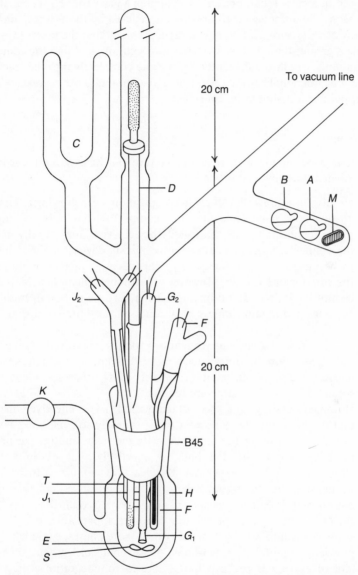

Fig. 3.27. The Pask–Plesch reaction calorimeter, approximately to scale. A phials of reagents, B phial magazine, C cold finger (not essential), D phial breaker, E vessel of calorimeter, F heater, G_1 Pt wires of the conductivity probe, G_2 terminals, H vacuum jacket, J_1 thermometer probe, J_2 terminals from thermometer probe and the compensating leads, K tap for evacuation of pseudo-Dewar space or admitting air, M magnetic pusher, T main tube, 20 mm o.d., $ca.$ 17 mm i.d., at the bottom of which one of the phials A will be when it is to be broken by the drop of D, S PTFE star stirrer.

introduced in a glass phial, broken by dropping a glass rod. However, there is no reason why ampoules with break-seals, as used by the Szwarc, Sigwalt and many other groups, should not be used, provided that the pressure in the ampoule is greater than that over the reacting solution, and that the contents of the ampoule can flow sufficiently rapidly and completely into the reaction mixture. For full details of the operation and advantages of this reactor, the original publication should be consulted.

3.3. Stirring

The agitation or mixing of a reaction mixture can have several purposes amongst which the obvious ones are:

(1) Homogenisation with respect to composition, i.e. achieving a mixture all of whose parts have the same composition: If only one liquid phase is concerned this can usually be achieved fairly rapidly unless the volume is large in relation to the mixing device, or the volumes to be mixed are very unequal, or the viscosity of one or more of the components is high. However, once homogeneity is achieved, it cannot be destroyed and mixing for the purpose of homogenisation becomes redundant, unless a product is precipitated during the reaction.

However, if a gaseous reagent or product is involved, the task of homogenisation becomes much more difficult, as the uptake and release of gases from solution are very slow processes. Generally, special designs of agitator are required.

(2) Homogenisation with respect to temperature: This effectively means that if conditions are to be adiabatic (or nearly so), the outer layers of the mixture which are in contact with the vessel and the gas phase must be mixed with the bulk so effectively that no important temperature difference persists. If a reaction mixture is to be kept isothermal, the stirring must be even more effective, and there is a danger that the stirrer mechanism may dissipate so much energy that the heat generated in the solution becomes appreciable.

(3) If one is dealing with systems of two or more phases, e.g. a catalytic powder in a liquid, or a two-liquid phase (emulsion) system, then the aim of stirring is to keep the surface-to-volume ratio as large as possible to maximise the rate of mass and heat transfer across the phase boundaries, and also to keep the composition of the mixture as uniform as possible, i.e. to prevent settling or creaming.

As far as vacuum systems are concerned, the stirring of reaction mixtures can be done by a reciprocating device or a rotating device, and the power can be transmitted to the moving part in the solution through a sealed rotating shaft (very rarely used with high vacuum) or by means of magnets. Colclough and Dainton used a reciprocating stirrer consisting of a glass-coated iron core inside a dilatometer, driven by a motorcar screen-wiper carrying a magnet (Colclough and Dainton, 1958). A reciprocating agitator operated by a solenoid and electronic 'flip-flop' was used by Longworth and Plesch for their study of the freezing-point phase diagrams of titanium tetrachloride with various alkyl chlorides (Longworth and Plesch, 1958), see Section 5.3.3.

By far the commonest form of stirring is by means of a rotor, and for our purposes only two methods of driving a rotor need be considered, both depending upon a rotating magnet. One mechanism consists of a long shaft, the lower end of which is fitted with a propeller and the upper with a glass-enclosed bar magnet or iron core. This rotates inside a glass tube fused to the reactor and is driven by a powerful horse-shoe magnet rotating outside the glass tube and above the iron bar, thus making up a magnetic clutch (Fig. 3.23). One can avoid much vibration in the reactor and its attachments if the rotating magnet is driven through a Bowden cable by an electric motor fixed firmly to a wall.

The crowded conditions inside a reactor due to the presence of various probes and a phial-breaker usually limit the number and size of propeller blades which can be accommodated. However, these fittings also break up the laminar flow (which is inimical to efficient mixing) so that turbulence can be achieved at stirring speeds well below that which would be required for a cylindrical reactor free of solid obstructions. If there is sufficient space, mixing can be improved considerably by having two propeller blades, *ca.* 1.5 cm apart, on the same shaft, with opposite chiralities, so that the layer of fluid between them is subjected to an exceptional shear-rate.

The other method available for stirring a closed reactor is by means of a free iron bar or bar magnet resting on the bottom of the reactor and driven by a rotating magnet below the reactor. The rotor is usually enclosed in glass or coated with PTFE and the commercially available stirrers are bar- or star-shaped. The advantages of the free rotor are that there is no driving shaft which takes up room and which can vibrate strongly. Its disadvantages are that the driving magnet must be fairly close to the reaction vessel and this restricts the size and shape of any thermostatting bath; also, the stirrer may get out of phase with the rotating magnet and jump around ineffectively and destructively; this is often caused by glass fragments entering the solution when a phial is broken, or by an abrupt change of viscosity which occurs when there is a sudden formation of a polymer.

When a new piece of apparatus is being designed, the need for, and method of, stirring must be given careful attention, because it is not feasible to give any very general guidelines on this subject.

114   Appliances and procedures

References

C. E. H. Bawn, C. Fitzsimmons, A. Ledwith, J. Penfold, D. C. Sherrington and J. A. Weightman, *Polymer*, **12**, 119 (1971).

V. Bertoli and P. H. Plesch, *J. Chem. Soc. (B)*, 1500 (1968).

D. N. Bhattacharyya, C. L. Lee, J. Smid and M. Szwarc, *J. Phys. Chem.*, **69**, 612 (1965).

R. H. Biddulph and P. H. Plesch, *Chem. and Ind.*, *(London)*, 1482 (1959).

R. H. Biddulph, P. H. Plesch and P. P. Rutherford, *J. Chem. Soc.*, 275 (1965).

G. J. Blake and D. D. Eley, *J. Chem. Soc.*, 7405 (1965).

S. Boileau, G. Champetier and P. Sigwalt, *Makromol. Chem.*, **69**, 180 (1963).

K. Brzezinska, K. Matyjaszewski and S. Penczek, *Makromol. Chem.*, **179**, 2387 (1978).

G. M. Burnett, *Mechanism of Polymer Reactions*, Interscience, New York, p. 15, 1954.

H. Cheradame, J. P. Vairon and P. Sigwalt, *Eur. Polym. J.*, **4**, 13 (1968).

R. O. Colclough and F. S. Dainton, *Trans. Faraday Soc.*, **54**, 886 (1958).

D. Dadley and A. G. Evans, *J. Chem. Soc. (B)*, 418 (1967).

A. G. Evans, D. Holden, P. H. Plesch, M. Polanyi, H. A. Skinner and M. A. Weinberger, *Nature*, **157**, 102 (1946).

J. C. Favier, M. Fontanille and P. Sigwalt, *Eur. Polym. J.*, **10**, 714 (1974).

S. L. Fries, E. S. Lewis and A. Weissberger, Eds. *Techniques of Organic Chemistry*, Wiley, New York, Vol. 8, Part 2, Chapter 21, 1963.

A. Gandini, P. Giusti, P. H. Plesch and P. H. Westermann, *Chem. and Ind. (London)*, 1225 (1965).

P. Giusti, *La Chimica e l'Industria (Milano)*, 435 (1965).

P. Giusti and F. Andruzzi, *La Chimica e l'Industria (Milano)*, 442 (1965).

M. Gordon, *Trans. Faraday Soc.*, **44**, 196 (1948).

M. Gordon and B. M. Grieveson, *J. Polym. Sci.*, **17**, 107 (1955).

D. W. Grattan and P. H. Plesch, *J. Chem. Soc. Dalton Trans.*, 1734 (1977).

M. A. Hamid, M. Nowakowska and P. H. Plesch, *Makromol. Chem.*, **132**, 1 (1970).

G. E. Holdcroft and P. H. Plesch, *J. Chem. Res.*, *(S)* 36 (1985); *(M)* 0401 (1985).

J. Jagur-Grodzinski, M. Feld, S. L. Yang and M. Szwarc, *J. Phys. Chem.*, **69**, 628 (1965).

F. R. Jones and P. H. Plesch, *J. Chem. Soc. Dalton Trans.*, 927 (1979).

N. Kalfoglou and M. Szwarc, *J. Phys. Chem.*, **72**, 2233 (1968).

C. P. Kesztelyi, Chapter 14 in *Laboratory Techniques in Electro-Analytical Chemistry*, Eds. P. T. Kissinger and W. R. Heinemann, Marcel Dekker, New York, 1984.

KPG means Keele Polymer Group (1951–85). The design is traditional and unpublished, or is based on one occurring in a Thesis.

B. Krummenacher, Dissertation No. 4778, ETH, Zurich, 1971.

B. Krummenacher and H. G. Elias, *Makromol. Chemie*, **150**, 271 (1971).

J. E. Lind, J. J. Zwolenik and R. M. Fuoss, *J. Amer. Chem. Soc.*, **81**, 1557 (1959).

W. R. Longworth and P. H. Plesch, *J. Chem. Soc.*, 451 (1958).

R. G. W. Norrish and K. E. Russell, *Nature*, **160**, 57 (1947).

W. Obrecht and P. H. Plesch, *Makromol. Chem.*, **182**, 1459 (1981).

S. D. Pask and P. H. Plesch, *Chem. and Ind.* (*London*), 331 (1981).

S. D. Pask, P. H. Plesch and M. DiMaina, *Chem. and Ind.*, (*London*) 329 (1981)

S. D. Pask and O. Nuyken, *Eur. Polym. J.* **19**, 159 (1983).

D. C. Pepper, *Trans. Faraday Soc.*, **45**, 404 (1949).

P. H. Plesch, M. Polanyi and H. A. Skinner, *J. Chem. Soc.*, 257 (1947)

P. H. Plesch, *Chem. and Ind.* (*London*), 699 (1973).

P. H. Plesch, the section on Dilatometry is a slightly altered version of an article in *International Laboratory*, October 1986, p. 18.

P. P. Rutherford, *Chem. and Ind.* (*London*), 1614 (1962)

C. D. Schmulbach and T. V. Oommen, *Analyt. Chem.*, **45**, 820 (1973).

H. A. Skinner, J. M. Sturtevant and S. Sunner, chapter 9, in *Experimental Thermochemistry*, Vol. II, Ed. H. A. Skinner, Interscience, New York, 1962.

V. Stannett, H. Garreau, C. C. Ma, J. M. Rooney and D. R. Squire, *J. Polym. Sci., Symp. Series*, **56**, 233 (1976).

M. Szwarc, *Carbanions, Living Polymers and Electron Transfer Processes*, Interscience, New York, Chapter 4, 1968.

D. J. Worsfold and S. Bywater, *J. Chem. Soc.*, 5234 (1960)

4. Purification, including drying

4.1. Preamble

One of the strongest reasons for using h.v.t. for any chemical reaction is to prevent access of interfering substances, which means maintaining a level of purity which has usually been achieved with considerable trouble. This justifies the present chapter. The subject is certainly not new, a general discussion having been given in the context of cationic polymerisation 26 years ago (Plesch, 1963). What was written there will be considered – as the lawyers put it – an integral part of the case here. As has been pointed out before, the question of purity is not of great concern to the preparative chemist as long as the impurity level is below, say, 1 mole %, but for those doing quantitative work it is a principal concern. In catalytic, photochemical, or electrochemical reactions impurities may act as co-catalysts, promoters, or inhibitors, thus affecting the rates of individual reaction steps. This may be reflected in the distribution ratios and stereo-chemistry of products, and in polymerisations by the molecular weight, the molecular weight distribution, and the nature of end-groups of the polymers.

One of the points made earlier (Plesch, 1963) is even more important today than previously, and that is the variability of the nature and quantity of impurities in any one compound with time and with its place of origin.

The changes with time stem from changes in manufacturing techniques and the progressive reduction in impurity levels arising partly from customers' demands and partly from legislation. An excellent example is provided by a set of analyses of several samples of MeCN obtained from various manufacturers (Dubois, 1982); see Table 4.1. It cannot be emphasised too strongly that this feature must always be kept in mind when reading older literature, especially when catalytic reactions are concerned.

The variability of quality between countries arises partly from differences of raw materials and of manufacturing methods from one country or one firm to another, from different commercial legislation regarding purity in different countries and, generally, from different standards of surveillance of quality in different countries. A good example of differences arising from manufacturing methods is found in methylene chloride; in the 1970s, some French workers could not obtain reliably reproducible results on certain polymerisations done in this solvent, despite heroic efforts at purification, as long as they used the material produced by a French refinery. When they followed the author's advice to use the BDH product, all was well. A similar situation is implicit in the fact that many researchers in the Comecon countries who are doing quantitative physico-chemical investigations spend their scarce hard currency in buying chemicals from Western suppliers.

As for the purification procedures to be adopted, these must be thought out as carefully as all other aspects of an experiment, perhaps even more so, as they are frequently very time-consuming. Each proposed procedure must be scrutinised as to its scientific basis, its appropriateness and its safety, and it is sensible to try it out on a small scale. The beginner especially must

Table 4.1. *Impurities in MeCN from different sources, concentrations in mole l^{-1}.*

Impurity	Merck	BDH	Fluka
MeOH	2.6×10^{-6}	2.4×10^{-5}	9.4×10^{-6}
EtOH	< d.l.[a]	1.9×10^{-4}	< d.l.[a]
EtCN	1.2×10^{-2}	6.2×10^{-3}	6.5×10^{-3}
$CH_2 : CHCN$	1.2×10^{-3}	1.5×10^{-4}	1.5×10^{-3}
NH_3	8.2×10^{-5}	9.4×10^{-4}	5.1×10^{-4}
$MeCO_2H$	6.7×10^{-2}	5.1×10^{-2}	3.7×10^{-2}
H_2O	1×10^{-2}	4×10^{-2}	2.8×10^{-2}

[a] Detection limit
MeCN from Prolabo, Carlo Erba and Aldrich gave analyses very similar to the results in the Table (Dubois, 1982).

beware of adopting uncritically any previously established procedures and should remember that most chemical practices can be improved; in particular, the novice must beware of performing meaningless traditional rituals.

One of the dangers which beset workers who do not use vacuum techniques can be safely ignored here, namely the deterioration of compounds by exposure to the atmosphere, such as hydrolysis, peroxidation, or the formation of carbonyl chloride from chlorine compounds by oxygen and light.

In the following sections we will indicate the general principles underlying different purification procedures, but we will not give details for the treatment of particular compounds, except as examples. An excellent short survey of many methods of purification has been given by McArdle and Sherwood (1985).

4.2. Definitions, relevance and measures of purity

4.2.1. Definitions and relevance
In a chemical context the concept of purity is circumscribed by three questions which need to be answered before much progress can be made:

(1) Purity of what?
(2) Purity for what?
(3) Purity from what?

Outline answers to each of these questions will be given here and detailed dicussions will follow in later sections.

(1) The brief answer to this question is 'everything': Apparatus, reagents, catalysts, solvents, adsorbents, drying agents, and (though strictly

not relevant here but worth noting) protective gases and gas streams.

(2) One must be clear about the purpose of the work, because for most preparative reactions in which yields greater than 90% are usually considered more than satisfactory, the level of purity required is relatively low. On the other hand, for quantitative investigations, especially kinetics and particularly those of catalytic processes, and many areas of electrochemistry and radiation chemistry, the concentration of impurities may need to be well below the milli-molar level.

(3) The answer is: From harmful substances, i.e. elements or compounds which can interfere with the processes to be studied. One may encounter compounds which neutralise a catalyst or which generate spurious catalysts, compounds whose intense fluorescence in minute concentrations may frustrate a photochemical experiment, or others whose adsorption as a monolayer may alter the behaviour of an electrode or some other surface.

The quest for purity is a campaign, and as in all campaigns the most important role is to 'know your enemy'; so, much of what follows will be concerned with the nature of harmful impurities arising from different sources.

4.2.2. Measures of purity.

Only three measures of impurity levels, as of concentrations in general, are generally useful: the molarity, the molality, and the mole-fraction or mole-percentage. Of these, molality is the least useful, the mole-fraction is rarely appropriate, and the molarity is to be preferred, as it is more informative and easier to use in calculations than the other two.

The terms 'p.p.m.' or parts per million and 'p.p.b.' (parts per billion) should never be used, as they are ambiguous and uninformative. *They are ambiguous*: Does p.p.m. mean by weight, or weight per unit volume, or is it in molar terms? *They are uninformative*: The significance of the p.p.m. number (if in weight terms) varies both with the molecular weight of the impurity and the density of the main constituent of the mixture. For a given p.p.m. number the corresponding molar concentration depends inversely on the molecular weight of the impurity. An impurity level of 10 p.p.m. by weight signifies purity of 99.999% by weight, and for water in benzene it means a concentration of *ca.* 4.4×10^{-4} mol l^{-1}, but for an impurity of molecular weight = 180 (instead of 18, as for water) 10 p.p.m. corresponds to 4.4×10^{-5} mol l^{-1}. And 10 p.p.m. of water in CCl$_4$ means 8.8×10^{-4} mol l^{-1}!

4.3. Cleaning the apparatus

The glass used for constructing the vacuum line and its ancillary equipment should have been cleaned conventionally and thoroughly at the start, but inevitably the hardware can become contaminated in various ways. The most common and ubiquitous contamination adhering to glass is undoubtedly water. Its importance can be shown by a simple calculation: Suppose that the surface of a 50 ml flask is covered with a monolayer of water; if this goes into solution in 50 ml of liquid it will give a *ca.* 10^{-5} mol l^{-1} solution of water, and actually the normal layer of adsorbed water on untreated glass is many molecules thick.

So what can be done about this? The water can be removed physically or chemically. The physical methods include washing the water off the surface or driving it off by heating. The washing is done by distilling a solvent continuously from a reservoir containing a drying agent through the apparatus and collecting the distillate back in the reservoir where the water, removed from the glassware by a partitioning process, is retained by the drying agent. As this method is much more appropraite to non-vacuum systems it will not be discussed further; it should be noted, however, that even polar solvents such as ketones do not displace the last layers of water.

Drying glassware by heating it under vacuum is very effective. For most purposes the glass is heated with a luminous brush flame until a pink glow appears in the flame. For the most rigorous drying, e.g. for studies on polymerisations initiated by high energy radiation, (Ueno, 1967) the whole reaction chamber was baked out just below the softening temperature of the glass. *Note*: Do not heat optical cells in this way.

The chemical removal of water can be done in several ways. For anionic polymerisations it is customary to rinse all the glassware repeatedly with a solution of sodium naphthalide which reacts with water to form naphthalene and NaOH, both of which are adequately soluble in the usual solvent tetrahydrofuran (THF) at the low concentrations used. This method, however, requires that the whole reactor assembly be detachable from the vacuum line so that it can be tilted around for the sodium naphtalide solution to reach all parts of the system. Care! If much water is to be removed and the volume of the purging solution is too small, some NaOH may be precipitated!).

An assembly which cannot be detached from the vacuum line can be freed from surface water by evacuating it thoroughly and admitting the vapour of trimethyl chlorosilane (see Fig. 3.11). The silane reacts with water to give trimethyl silanol and HCl, and the OH-groups of the glass surface are converted to Me_3SiO-groups, so that the inside of the apparatus is effectively covered with a layer of methyl groups, i.e. paraffin wax. The HCl evolved must be pumped off very thoroughly. A glass surface treated in this way will retain its 'waxy' layer unless it is washed with very aggressive reagents, e.g. KOH + aqu. ethanol or is put through an annealing cycle (see p. 20), and adsorbed water can be pumped off from it fairly easily.

Another way of ensuring that a glass surface is free of water or hydroxyl groups is to cover it with a film of sodium. This method was used when the melting point phase diagram for the system isobutene + titanium tetra-chloride was determined (Longworth, Plesch and Rutherford, 1959; Plesch, 1972), but it cannot be used on silane-treated glass as the sodium will not spread on the waxy surface.

4.4. The cleaning of solvents and reagents

4.4.1. General introduction

A more general heading could have referred simply to 'organic compounds'; the reason for treating solvents and reagents together is that one chemist's solvent may be another's reagent, e.g. THF or vinyl chloride, both of which are good and interesting solvents, as well as yielding lucrative and interesting polymers. The principles of purification are very much the same, whatever the ultimate use of a compound. However, the purification of any substance to be used as a solvent is more critical than that of a reagent because generally, and especially in quantitative work, the solvent is much the more abundant species. Consider a reaction mixture in which a reagent R is present at a concentration of 0.1 mol l^{-1} and the concentration of a catalyst C is 10^{-3} mol l^{-1}. If the molecular weight of R is 100 g mol^{-1} and its density 1 g ml^{-1} and it contains 10^{-2} mol l^{-1} of an impurity Imp, then the 10 g of R in 1 l of mixture introduces a [Imp] = 10^{-4} mol l^{-1}, i.e. if Imp reacts with C, it will neutralise one tenth of C. If the solvent contains the same concentration of the same or another noxious impurity, it would inactivate all of C, swamping it by a factor of 10. In fact the maximum tolerable [Imp] in the solvent would be 10^{-4} mol l^{-1}, i.e. one hundredth of that in R. These are primitive considerations but they are worth keeping in mind.

This is not the place for a collection of purification recipes, as enough of these are available (see, for example, Coetzee, 1982; Perrin, Armarego and Perrin 1983), but it is appropriate to gather together the principles of various purification techniques, because the last and most critical stages are likely to be carried out in a vacuum system.

On the subject of 'know your enemy', one must think of two classes of impurities: (1) Those of chemical type similar to the principal compound, e.g. a higher alkene in propene, or 2-butanol in 1-butanol; and (2) those of different chemical type, possibly derived from the main compound by oxidation or degradation, e.g. a carboxylic acid in an aldehyde or an N-oxide in an amine. In this context it is worth noting that if one is concerned with reactions sensitive to trace impurities, the organic chemist is a poor guide to purity matters, because in his preparative activities he is generally not concerned whether, for example, a few tenths of a per cent of an alcohol being used as a reagent or solvent has been dehydrated to an alkene, or whether a cyclic formal contains traces of peroxides or formaldehyde, whereas to the kineticist or electrochemist such impurities, even in traces, may be critical.

The process of purifying a compound generally comprises several stages, all but the last two or one of which are carried out before the compound is introduced into the vacuum system and which therefore do not concern us here. It should be remembered, however, to design all stages of the procedure so that no new noxious impurity is introduced which can then be carried along through the succeeding stages. For example, it is unwise to crystallise a compound to be used in photochemical reactions from an aromatic solvent or a ketone, or to crystallise from an acidic compound (in the widest sense) a monomer which is to be polymerised anionically.

4.4.2. Physical methods

Although all the physical methods of purification, which comprise the various forms of distillation and of crystallisation, as well as adsorption, can be done in a high vacuum system, the only ones for which this is generally worthwhile are crystallisation without solvent and its sophisticated version zone-refining, and adsorption. Of course, degassing cannot be done otherwise than on a vacuum system.

4.4.2.1. Degassing. When any liquid is introduced into a vacuum system it contains the equilibrium concentration of the gaseous constituents of the atmosphere and before anything can be done with this liquid these impurities must be removed. The traditional method of doing this is by so-called freeze–pump–thaw cycles. The liquid is frozen, the space above it is pumped out and the liquid is thawed. Thereupon the dissolved gases partition themselves between the liquid and the space above it. As was mentioned earlier, the partitioning of gases between liquids and the gas phase is a very slow process, which can be accelerated by vigorous stirring. When the liquid ceases to bubble, it is refrozen, pumped, and thawed again. This procedure is repeated until the pumping is no longer accompanied by a change of tone of the rotary pump. With liquids of low vapour pressure it can be very efficient to pump cautiously for a few seconds before the freezing starts. Evidently, the larger the volume into which the vapour of the liquid plus the dissolved gases can expand, the more efficient each cycle will be.

The freezing should always be done in a Kon flask (inverted cone) which should not be filled above the 'equator'. If these precautions are observed, the frozen cake, when it expands upon being warmed, can simply slide upwards and will not crack the flask. It is generally not efficient to degas volumes greater than *ca.* 200 ml. For larger volumes, usually stored in round-bottomed flasks, the degassing must be done without freezing. In this case, to avoid exposing the liquid to the pumps directly, it is best to let the vapour and gases expand into a ballast volume roughly equal to that of the liquid, closing a tap between the liquid reservoir and the ballast vessel, and then pumping out the latter so slowly that the condensable vapour is held completely by the cold traps.

4.4.2.2. Crystallisation from the melt.   Crystallisation from a liquid which consists mainly of a single compound is a very effective method of purification. It is based on the phase behaviour of a binary mixture, as illustrated in the binary freezing point phase diagram, all the impurities being considered as a single 'second component'. It is usually applied to compounds with melting points near ambient temperature, such as t-butanol or acetic acid, by cooling the liquid slowly whilst stirring it or rotating the container about a horizontal axis. As the main component crystallises, the impurities accumulate in the supernatant liquid. Usually, this is syphoned or filtered off, the crystalline material is melted and refrozen, and the procedure is repeated as often as may be required; the fraction of unfrozen material which is discarded at each stage can be reduced progressively. Provided that none of the impurities involved form solid solutions with the main component in the concentration range of interest, this method is extremely efficient. It can be used on a vacuum line without difficulty, but only on relatively small volumes, say up to 200 ml. The advantage of handling small volumes is that a much larger temperature range is accessible for purely practical reasons. The principal consideration is that the cooling be done very slowly and evenly.

4.4.2.3. Zone-refining.   The purification process known as zone-refining is an elaboration of the crystallisation without solvent described in the previous section (McArdle and Sherwood, 1985). The material to be purified is melted into a tube, generally from 0.3 to 1.3 cm diameter and *ca.* 20 cm long. This is passed upward through one or more narrow annular heated zones in which the material melts. As the molten material passes upward out of the heated zone, the purer material freezes first and impurities accumulate in the melt which finishes up at the bottom of the tube after this has passed through the hot zone(s). The tube is then returned to the starting position and the process is repeated. In this way the impurities are swept down toward the end of the tube and the material at the top of the molten ingot is the purest. Under normal operating conditions the whole ingot can be recovered by cutting up the tube into appropriate lengths and melting the contents of each segment. A big advantage of the method is that the impurities are trapped at the end of the ingot and can be analysed.

In some systems an impurity may partition itself in such a way that it is swept to the top of the ingot, and of course both types of impurity may occur in the same material. In such cases only the middle part of the ingot has the required purity. (The author thanks Prof. J. N. Sherwood for some personal advice on this point.)

Some compounds expand so much during the melting and freezing cycles

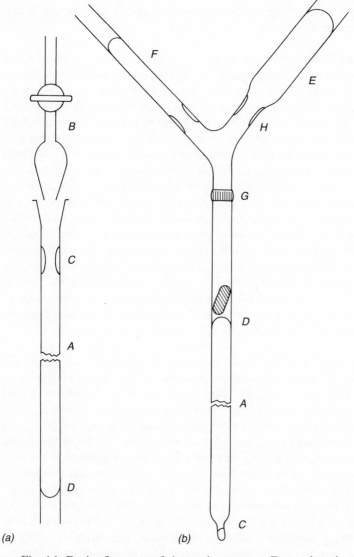

Fig. 4.1. Device for zone-refining under vacuum. For explanation see text.

that they break glass tubes. This mishap can be prevented if one places a thin
tube of polyethylene or some other inert plastic, sealed at the lower end, into
the zone-refining tube. The deformability of such a tube suffices to absorb
the stresses generated by the expanding ingot. When attempting to use the
method and when assessing its efficiency in any particular case, one must
remember that its success too depends upon the material behaving as a

normal binary mixture with a single eutectic, at least in the composition region relevant to the purification (as was discussed in the previous section). If this is not so, i.e. if the main component forms a solid solution with one or more of the impurities, then these cannot be removed effectively by zone-refining.

There is no difficulty about evacuating a zone-refining tube after it has been charged and the contents melted. It can then be detached from the vacuum line after a small tap at its apex has been closed or after it has been sealed off; but the transfer of the desirable, upper part of the zone-refined ingot to the vacuum line requires a special development of the standard zone-refining tube. The author's design is shown in Fig. 4.1. and is used thus: As shown in (a), tube A is filled with the material which is melted so as to leave as few voids as possible. Whilst still molten, it is evacuated through the tap-stopper B, and is left to solidify whilst being evacuated. When the ingot has cooled, A is fused off at C and the material is zone-refined with the seal-off point C at the bottom. When the zone-refining has been completed, the tube is fused to two (or more) break-seal ampoules E and F, as shown in (b). Seal D is broken, the wanted part of the ingot in A is melted and poured into ampoule E through the coarse sintered filter G. By judicious use of a small flame the seal-off point H is freed from adhering product and then sealed off. The ampoule E containing the purified compound is then ready for use. The residue in A containing the impurities can be transferred to the second break-seal ampoule F for analysis, or is discarded. Evidently, more than two break-seal ampoules can be used, and there is little difficulty in discarding the top part of an ingot and collecting only the middle part for use.

4.2.4. Adsorption.   The process of adsorption involves the partitioning of the adsorbed substance between the bulk phase in which it is dissolved and the surface of the adsorbant, and it is a reversible, equilibrium process, unless the adsorbed species is transformed or bound chemically; the use of solid reagents to trap impurities by chemical reaction will be discussed in the following section.

The most common adsorbants are various forms of alumina, silica and molecular sieves. The adsorbates which are removed preferentially from an organic bulk phase are more polar, or polarisable, than the compound constituting the bulk phase. One must distinguish between the *adsorptive capacity* and the *adsorptive strength* of an adsorbant. The adsorptive capacity depends upon the available surface area i.e. the specific surface ($cm^2\ g^{-1}$), whereas the adsorptive strength is the binding energy of the adsorbate on to the adsorbant, and this depends, of course, on the nature of the adsorbate for any one adsorbant. These two quantities together determine the residual concentration of the adsorbate in the bulk phase for any given set of conditions.

It is generally found that the best adsorbants have a relatively small adsorptive capacity but a very high adsorptive strength. That is why it is advisable to reduce the concentration of impurities by some pre-treatment with an efficient reagent of high capacity, e.g. fuming sulphuric acid to remove alkenes, aromatics, alcohols, etc. from a saturated hydrocarbon. Thereafter the bulk phase can be treated on a high vacuum system with an agent of low capacity (since there is now a low concentration of impurities) but of high adsorptive stength, to give ultimately a very low equilibrium concentration of impurities. These considerations govern all kinds of systems, but they are especially important for the removal of water. If the last few paragraphs are reread with the words 'water' and 'drying' substituted in the appropriate places, they will give an adequate introduction to the theory of drying. Some special explanations about drying will be found in Section 4.6.

4.4.3. Chemical methods

Chemical purification means the transformation of the unwanted contaminants into different chemical species, preferably firmly bound to a solid phase, but in any case (virtually) insoluble in the wanted material. Several chemical methods are pre-eminently suitable for h.v.t. and most of these involve removal of water. It must be remembered, however, that many of the reagents concerned will react not only with water, but with any compound having an active hydrogen, such as alcohols. The use of Na + K alloy is described in Section 4.6.2.

4.4.3.1. Sodium.   Probably the best known 'active hydrogen' remover is sodium. When used outside a vacuum system, for instance as sodium wire to dry solvents, the sodium is little more than a support for a skin of sodium hydroxide. Inside a vacuum system, however, one can prepare films of sodium metal and one can prepare really clean sodium which will give a colourless solution of sodium ethoxide (see Section 5.2.1.). The method of making sodium films for the removal of 'acidic' compounds from liquid reagents will be described and also a very much less well-known method involving sodium vapour and colloidal sodium.

Formation and use of sodium films.   Fig. 4.2. shows the apparatus used to prepare sodium films. A lump of sodium is cut from stock and the adhering oil removed with filter paper and a quick rinse in hexane. It is then placed in the cup A which is warmed gently with a luminous flame until the sodium melts into the container B leaving its skin C behind in A; thereupon the apparatus is sealed at D, care being taken that none of the sodium skin is fused into the glass. If that happens the seal is likely to crack on cooling and therefore the preparation must be abandoned by sealing off at F. If all goes well, vacuum is applied as soon as the seal-off point has cooled. When the vacuum is satisfactory, the sodium in B is heated with a free flame and a

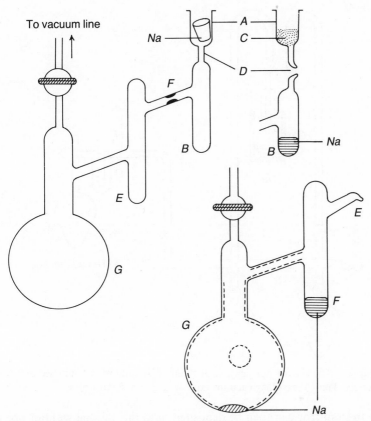

Fig. 4.2. Assembly for making sodium mirrors. The method is explained in the text. The same apparatus can be used for making P_2O_5 'snow'.

substantial part of it is distilled into E, a black residue remaining in B, which is then sealed off at F and discarded. (Take the prescribed precautions when disposing of waste sodium!) During the distillation the connecting duct must be kept warm by means of the flame to prevent it becoming blocked by frozen sodium. Next, a small amount of sodium is distilled into G. During this operation a sodium mirror (dotted lining) will begin to form on the walls of G. More sodium is then deposited there by heating the blob of sodium at the bottom of G briefly with a fairly hot, small flame; the aim is to evaporate the metal with minimal heating of the area in which it is to be deposited. When this operation is completed and the inner surface of G is covered with an even metallic deposit, a small flame is applied briefly to clear a window (dotted circle) through which the state of the surface of the sodium can be observed. Such mirrors are used very frequently to dry solvents or reagents. When after some time the surface becomes dull, the vessel is pumped out very thoroughly

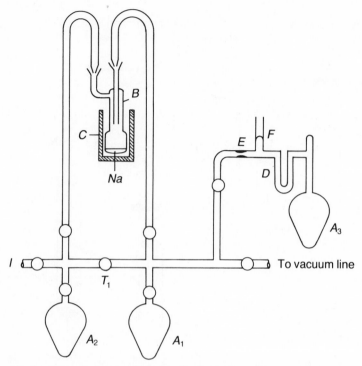

Fig. 4.3. The rig for purification of a liquid by sodium vapour. See text for details. The ○ are glass vacuum taps of large (> 5 mm) bore.

and a fresh mirror can then be deposited onto the old one without opening the system to the atmosphere; the other alkali metals (but not lithium) can be handled by analogous techniques.

The use of sodium vapour. A very effective method of removing any impurities which react with sodium (including dienes) involves sodium vapour and subsequently colloidal sodium in the rig shown in Fig. 4.3. This was devised in the 1950s (KPG), but was first published by Nuyken (1981), for purifying isobutene.

Gaseous isobutene from a cylinder is condensed through the inlet I, into the Kon flask A_1, degassed, and distilled slowly into A_2, with T_1 closed, through trap B. This contains sodium which is heated by the furnace or mantle C to *ca.* 300 °C. The process is then reversed and the isobutene is distilled back into A_1, this time rather more rapidly so that it carries some sodium vapour with it which condenses to colloidal sodium, which fairly soon turns from purple to grey. The process is repeated, and this time the purple colour will last rather longer. Finally, the isobutene is distilled very rapidly into A_3 so that it carries a lot of sodium vapour with it. A freeze seal is made at D and the reservoir A_3 is sealed off at E. The isobutene can then be stored until needed, when it will be fused to the apparatus via the break-seal F. Since at ambient temperature the pressure in A_3 will be *ca*. 5 atm, it should be bandaged with adhesive tape and kept in a wire cage. In order to obtain fairly rapid distillation and minimum obstruction, all the taps should be large-bore glass taps, lubricated by a well-dried hydro-carbon grease.

4.4.3.2. Phosphorus pentoxide.   In order to obtain really pure and effective P_2O_5 one can use the apparatus shown in Fig. 4.2, but with a slight modification. There is no need for the constriction D, as it is not possible to prevent the glass being covered with P_2O_5, and one cannot make a satisfactory seal with glass which is contaminated in this way. Therefore the P_2O_5 is shovelled with a spatula from the stock bottle into B through A, the inside of the cup A is wiped clean and A is plugged with a rubber bung. The whole apparatus is then evacuated: cup A must be conical and of an angle similar to that of the rubber bung so that the sucking in of the bung by the vacuum does not split the cup. The P_2O_5 is then sublimed through F into E (and some of it will travel to G), leaving behind the involatile hydration products. Seal-off point F is then cleared by warming it, and it can then be sealed off without the inclusion of any P_2O_5 in the fused glass. Next, the tap is closed and the P_2O_5 in E is heated with a brush flame, whereupon it volatilises and comes down in G like a blizzard of minute white flakes; this is now a really effective reagent. However, it is not pure, because P_2O_5 usually contains some P_2O_3 which is also volatile; both compounds have a small, but noticeable solubility in many organic compounds.

4.4.3.3. The pre-treatment method.   An old and sound and very general method of purification, especially relevant to solvents, is to expose the compound in question to the action of the agent which is to be studied in that solvent. If the solvent is to be used for electrochemistry it should be electro-lysed with a DC potential gradient of $50-100$ V cm^{-1} until the resistance becomes too high for a significant current to pass. If it is to be used for photochemistry, it should be exposed to the radiation to be used in the experiment. If it is to be the medium for a catalytic reaction, it should be exposed to the action of the appropriate catalyst so that this may attack (and, one hopes, remove) the impurities which are sensitive to it. This method has been much used in the context of all kinds of polymerisations and in many kinds of physico-chemical work (Fairbrother, Scott and Prophet, 1956). A thorough study of the progressive improvements in the quality of a solvent which can be made in this way was reported by Grattan and Plesch (1977). They wanted to obtain stable solutions of $AlBr_3$ in MeBr in order to study their electrochemistry, but found, like most previous investigators, that when $AlBr_3$ is dissolved in MeBr there are irreproducible, time-dependent, changes in colour and electrical conductivity. Thereupon they subjected the solvent to

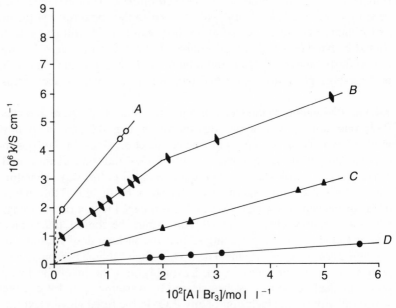

Fig. 4.4. The conductivity of AlBr$_3$ in MeBr at $-64\,°C$ as a function of concentration and of the successive purification cycles (see text). The conductivity of the solvent without AlBr$_3$ was: A, $\kappa = 9.4 \times 10^{-8}\ S\ cm^{-1}$; B, $\kappa = 4.9 \times 10^{-8}\ S\ cm^{-1}$; C, $\kappa = 1.9 \times 10^{-8}\ S\ cm^{-1}$; D, $\kappa = 0.34 \times 10^{-8}\ S\ cm^{-1}$.

a large number of cycles consisting of alternating contact with a sodium mirror and with a fresh sample of AlBr$_3$. After every two or three cycles the dependence of the electrical conductivity on the concentration of AlBr$_3$ was determined (Fig. 4.4). This was continued until a straight line through the origin was obtained (line D) which did not change with further purification cycles or with time; from the characteristics of the system one could calculate that at this stage the concentration of interfering impurities was less than 10^{-6} mol l^{-1}.

One lesson, which actually should be fairly obvious, is that it is not worthwhile to subject any compound to a repetitious purification routine unless one has an adequate means of monitoring the progress of the purification.

4.4.4. Purification of catalysts

4.4.4.1. General considerations. In the present context the term 'catalyst' comprises inorganic, metal-organic, and organic compounds, some of which are catalysts in Berzelius' original sense, i.e. they are not consumed during the reactions which they catalyse, and others which should be termed 'initiators', i.e. compounds which are wholly or partly incorporated in the reaction products; some of the co-catalysts and promoters encountered in catalytic systems are also included. We confine our attention entirely to compounds soluble in the reaction medium and exclude solid, insoluble catalysts such as those used for hydrogenation, cracking, etc.

Because of the immense variety of catalysts, it is difficult to give any generally applicable advice and suggestions for their purification and use in high vacuum systems. However, before entering upon the subject, it is worth examining whether extreme purification of a catalyst could be a worthwhile undertaking. The reasoning behind the question is that the concentration of the catalyst is usually so small that any impurity present to the extent of, say 1 mole % would not be expected to have any significant influence. In fact, there are several instances where a putative catalyst or co-catalyst was, itself, not active, but its apparent catalytic activity was due to a highly active impurity. Perhaps the most widely known instance relates to the use of boron trifluoride. Ever since the classic experiment of Evans and Meadows (1950) in which it was shown that BF_3 alone does not initiate the polymerisation of isobutene, and the studies of Eastham (Eastham 1956; Eastham and Clayton, 1957) on the interaction of BF_3 with water, the need for a co-catalyst was understood. However, the nature of the active co-catalyst in different systems was obscure; many workers believed that protons from residual water were the true initiators, even when BF_3OEt_2 was used, whilst others supposed that in such reactions the diethyl ether provided an initiator in the form of the Et^+ ion. The issue remained undecided until it was shown by means of the 'proton sponge' 2,6-di-t-butyl pyridine that in such systems not Et^+, but protons (presumed to originate from traces of water) are the true initiators (Moulis, Collomb, Gandini and Cheradame, 1981; Gandini and Martinez, 1987). If someone had undertaken to purify the BF_3OEt_2 adequately, or had synthesised it from adequately pure components in a high vacuum system, the question could have been settled long ago.

Another very instructive case concerns the alleged initiation of a cationic polymerisation by a charge-transfer complex formed by the compound chloranil (2,3,5,6-tetrachloroquinone) with the monomer N-vinyl-carbazole. It was shown (Natsuume et al., 1969; 1970) that this compound is not an initiator, but that the polymerisations were caused by a hydrolysis product, 2-hydroxy-3,5,6-trichloroquinone, which is a strong acid. One has learnt from this finding to be extremely suspicious of any claims for charge-transfer catalysis and to test one's suspicions by appropriate experiments involving progressive purification of the putative catalyst.

A very important category of 'mistaken identity' concerns the polymerisations allegedly initiated by acetyl perchlorate. Unless such experiments are conducted in high vacuum systems with really good technique, the residual water to be found under almost all other conditions produces a mixture of acetic and perchloric acids. The $HClO_4$ will participate in the reaction together with any unhydrolysed acetyl perchlorate. The real trouble is that since the amount of water in the systems is not known, the exact amount of perchloric acid formed, and therefore the true concentrations of the initiators are unknown in all such systems. The polymerisation results obtained by Higashimura and his school and by many others with $MeCO^+ClO_4^-$ and similar compounds are to be interpreted with this in mind.

To conclude, it is worth recording the advice given to the author at the very start of his career by the veteran catalytic chemist Alwin Mittasch, who had been Fritz Haber's 'officer in charge' of catalyst research for the ammonia synthesis: 'In all catalytic studies only the very purest is good enough'.

4.4.4.2. Simple compounds. The best way of obtaining many simple inorganic compounds in the required state of purity is to synthesise them from the elements. Many of these elements can now be obtained extremely pure, and for very many such syntheses the h.v.t. is the method of choice. As for the selection of an appropriate synthetic reaction, if the standard sources, such as *Inorganic Syntheses*, do not provide a convenient method, it is worthwhile to search the older literature for the syntheses used by the chemists who were working on the determination of atomic weights before the advent of the mass spectrometer. Some examples of the synthesis of metal halides will be given in Chapter 5. Because of the aggressive nature of these catalytic metal halides, most attempts at purifying them by crystallisation or any other method involving a solvent (even SO_2) generally introduce more impurities than they remove, because the solvent must be extremely pure if it is not to contain more contaminants than the metal halide. This is an instance where the pre-treatment of the solvent with the compound to be purified is essential for success. If the presence of impurities is suspected, purification methods without a solvent, such as fractional crystallisation (Section 4.4.2.2.) or zone-refining (section 4.4.2.3.) are preferable.

4.4.4.3. Crystallisable salts and related compounds. Almost all crystallisable catalysts, such as sodium and lithium aromatic compounds (e.g. sodium naphthalide), -oyl salts such as aroyl hexafluorophosphates, alkoxides, and many others can be prepared in a vacuum system and then purified by repeated crystallisations and washings in a closed system (see Chapter 5); thereafter they can be distributed into breakable phials or other devices as described in Chapter 3.

4.4.4.4. Acids. Most acids of interest in the present context except the halogen acids are liquids which are difficult to crystallise, and most have such high boiling points that even when distillation under vacuum is attempted, it is likely that some decomposition occurs, so that the final product may be more impure than the initial material.

One method which is little known, but very useful for purifying intractable liquids, is fractional precipitation; it is an equilibrium method and is a form of liquid–liquid extraction.

A necessary preface to a description of the procedure is that the solvent and the precipitant must be purified to exhaustion by contact with successive specimens of the acid to be purified. The acid A is dissolved in the minimum amount of solvent S. The precipitant P is then added under isothermal conditions to the solution until roughly one half to three quarters of A has been precipitated. At this stage there is a three-phase system present (vapour and two liquids) with three (or more) components (A, S, and Imp where Imp denotes an impurity), and the impurities are partitioned between A and the mixture of S and P. This mixture is separated from A by decantation or syphoning, A is redissolved in S and reprecipitated by the addition of P. At all stages of this process the mixtures must be stirred efficiently but so gently that an emulsion is not formed. It happens quite often that an acid A with a melting point near or above ambient temperature will start to crystallise after the first or second extraction.

The whole procedure can be done on a vacuum line, but is certainly easier and swifter in the open; if done 'on the bench', one must take precautions against the acid (or indeed any other liquid) picking up atmospheric moisture during the processing. (Inert gas blanket or dry-box, p. 5)

The most useful solvents are diethyl ether and acetone, and pentane and cyclohexane are amongst the best precipitants, i.e. worst solvents. It is not widely known that amongst hydrocarbons the lower the molar mass, the worse are the solvent properties, and there is a distinct difference in this respect between, say, pentane and heptane.

4.5. Determination of purity

4.5.1. Introduction

The methods for determining chemical purity fall into several categories:

(1) Detection of impurities and their estimation without separation, e.g. by their effect on colligative properties, or by some kind of spectro-scopy, or an electroanalytical technique such as polarography.

(2) Detection of impurities by separation, which almost always involves the finding of unexpected peaks in some kind of chromatogram, and which may lead to identification.

(3) Detection and estimation of impurities by some kind of functional test.

All tests need to be assessed and the testing of tests is such an obviously important procedure that its mention may seem superfluorous. It is, however, especially important in the present context, because if one is concerned with purity in relation to catalytic reactions, the domain of interest is, roughly, from 10^{-2} mol l^{-1} downwards, and with regard to many analytical methods one is going into unknown territory beyond the usual scope of the instruments.

The most useful method of testing the efficiency of a test is by means of a conventional calibration in which the response of the instrument is plotted against the concentration of a compound selected as a suitable model for the suspected impurity (if the nature of this is not known), with the main compound being used as the solvent. The concentration range should be carried down as far as possible to ascertain the lowest detection limit for the impurity.

4.5.2. Effects on physical properties

The effect of impurities on the transition temperatures of a compound (which in practice means the melting point or the triple point, see Chapter 1) gives the total molar concentration of impurities. It is not a generally useful criterion of purity in the domain of interest here, as the following considerations will show. When an organic chemist is satisfied with a melting point this means usually an agreement to ± 0.1 K with the 'accepted' value. Suppose we have an impurity of molecular weight 100 and a main compound which gives a melting point depression $\Delta T = 100$ K for 1 g mol of solute in 100 ml. This means that $\Delta T = 0.1$ K is equivalent to 10^{-2} mol l^{-1} of impurity. Since we are interested in impurity levels from 10^{-2} mol l^{-1} downwards, which implies $\Delta T \ll 0.1$ K, it would take elaborate and expensive equipment to monitor the progress of any purification that is of interest here.

A very comprehensive treatment of thermal methods for determining purity has been given by Wendlandt, but not much of it is of direct relevance to the matters discussed here (Wendlandt, 1986).

Both IR and NMR spectroscopy are useless for the detection and estimation of impurities at the levels of interest in the present context, although they may serve to identify impurities once they have been accumulated, as in the bottom residue in a zone-refining tube.

The UV spectrum is also of very limited use, because for any hope of success, the spectrum of the impurity must have an absorption peak in a region in which the main compound absorbs only feebly; evidently, aromatic impurities in aliphatic compounds can be detected easily in this way down to very low concentrations, as can polycyclics in mono-cyclic aromatic compounds.

This section would be incomplete without a mention of fluorescence spectroscopy, because most compounds having an UV-excitable fluorescence can be detected in extremely low concentrations, well within the range of interest here.

4.5.3. Mass spectrometry

The technique of mass spectrometry (MS) is itself of no help in the detection of impurities, but in combination with gas chromatography (the GC-MS technique) it would be the method of choice for the identification of impurities.

4.5.4. Electroanalytical methods

Without doubt the various forms of polarography are among the most generally useful analytical techniques because of their great sensitivity and the relative simplicity and cheapness of the equipment. There are many publications dealing with this topic. The conditions for using it successfully for analysing organic compounds are: (*a*) that a suitable solvent can be found, (*b*) that the impurities can be reduced at the dropping mercury – or any other suitable – electrode, and (*c*) that its half-wave potential should be sufficiently distant from that of the main compound.

In order to analyse, by polarography compounds which have been purified in a h.v.s., it is not necessary to have a cell which operates under vacuum. Methods have been described whereby a solution made up under vacuum can be admitted via a burette fitted with a pressure equaliser to a polarography cell operating at atmospheric pressure under nitrogen or, preferably, under argon (Kabir-ud-Din and Plesch, 1978). See also Chapter 3.

4.5.5. Detection and estimation of impurities by separation

The detection, identification and estimation of impurities by various chromatographic techniques is so well documented that few comments are required here. It must be remembered, however, that since we are concerned with extremely low concentrations, one cannot be sure of finding an impurity whose retention time on a column is close to that of the main compound, and also that a very small fraction of the main compound may undergo transformations on the column, especially if it is at an elevated temperature, so that spurious impurities may be produced in this way. If the main compound is sensitive to any of the components of air, especial precautions must be taken in transferring the sample from its evacuated container to the inlet of the chromatograph.

One advantage of gas chromatography is the availability of detectors which respond specifically to certain types of compound. The best known are the electron capture detector for chlorine compounds and the flame photometric detector for nitrogen and phosphorus compounds. If one wants to detect very small molecules such as water or CS_2, the standard flame ionisation detector must be replaced by a thermal conductivity detector.

4.5.6. Functional tests of purity

The most logical way of testing the purity of a reagent or a solvent is by means of the reaction in which it is involved. One of the best examples is that illustrated in Section 4.4.3.3. and Fig. 4.4, where it is shown that the purity of a solvent can be checked very effectively by equilibrium measurements, namely the shape of the curve representing the dependence of the electrical conductivity of a solution on the concentration of an ionogenic solute. However, this test only tells one whether the purity is adequate; it does not give information on the nature of the impurities nor on their final concentration, although a rough estimate of this can be made on the basis of the concentration of ions in the final solution.

The effects of progressive purification of both the solvent and the solute used in spectroscopic studies can be followed similarly in terms of changes in the spectrum. If the spectral characteristics of the solute tend to an asymptote as purification proceeds, one may conclude that interfering impurities are being removed progressively.

Many investigators faced with problems of purification do not realise that in many instances the impurities themselves are not discernible by the measurements being used in the experiments, but that they can produce interference by reacting with one of the reagents. For example, the studies on the system $AlBr_3 + MeBr$ quoted in Section 4.4.3.3. showed that during the later stages of the purification the conductivity of the MeBr was negligible compared to that of the solutions of $AlBr_3$ in less pure MeBr. When $AlBr_3$ was added to this highly, but not exhaustively, purified MeBr, its conductivity became much higher than that of a solution of the same concentration in exhaustively purified MeBr. This indicates the presence of an impurity which reacted with the $AlBr_3$ to produce ions.

An example of a kinetic test for purity is provided by experiments on the rate of polymerisation of N-vinylcarbazole (NVC) by mercuric chloride in chlorobenzene illustrated in Fig. 4.5. (Hamid, Nowakowska and Plesch, 1970). Curve A was obtained from kinetic measurements on NVC I which had been thrice crystallised from n-hexane to give a yield of 70%. The fact that the first-order rate constant k_1 of the polymerisation becomes independent of [NVC] only when the [NVC] exceeds $ca.$ 0.25 mol l^{-1} indicates that the monomer contains a co-catalyst. If one assumes that this reacts on a 1:1 basis with the $HgCl_2$ whose concentration was 8.33×10^{-4} mol l^{-1}, one finds that the co-catalyst content of the NVC was $ca.$ 1.3 mole %. A further two crystallisations with yields of 50% to give NVC II approximately halved the co-catalyst concentration (curve B). Two zone-refinings gave NVC III (curve C), and two further zone-refinings gave the single point D, which indicates that the limit of purity attainable by these methods had probably been reached, but also that the monomer probably still contained some co-catalytic impurity.

A different method of determining the concentration of co-catalytic impurities in a different polymerisation system was used in studies of the polymerisation of isobutene (Biddulph, Plesch and Rutherford, 1965) and of styrene (Longworth, Panton and Plesch, 1965) by $TiCl_4$ in CH_2Cl_2 in which

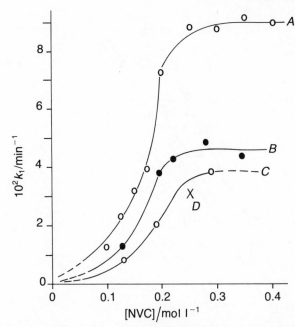

Fig. 4.5. The effects of progressive purification. The figure shows the first-order rate constant k_1 for the polymerisation of N-vinylcarbazole (NVC) as a function of the [NVC] for NVC purified in different ways; for details see text.

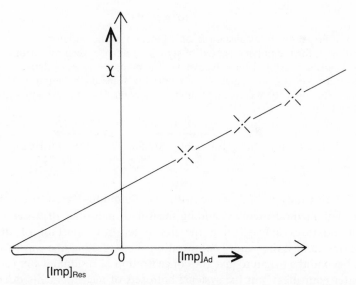

Fig. 4.6. Method for determining the concentration of residual impurities, $[Imp]_{Res}$, in a reaction mixture if the impurity is a catalyst or co-catalyst. The observed variable χ can be peak-height h for a GLC method, absorbance A for spectroscopy, conductivity κ for conductimetry, current i for polarography, or rate constant k for kinetics, etc.

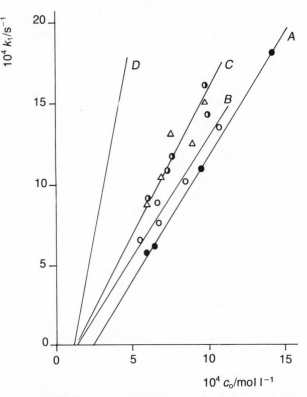

Fig. 4.7. The first-order rate constant k_1 as a function of the initiator concentration c_0 for the polymerisation of styrene in nitrobenzene by various salts; for details see text and the table below. The pumping time, t_p, determines the concentration of residual water which neutralises some of the initiator. In this instance the impurity acts as an 'inhibitor', in contrast to the examples in Figs. 4.5 and 4.6

Line	A	B	C		D	
Initiator	EtCOSbF$_6$	EtCOSbF$_6$	EtCOSbF$_6$	NO$_2$SbF$_6$	EtCOPF$_6$	Ph$_2$HCSbF$_6$
t_p/h	1	4	8	8	16–24	

water acts as co-catalyst, both done with the Biddulph–Plesch reaction calorimeter. The method involved adding measured quantities of water to the reaction mixtures and plotting the first-order rate constant of the polymerisations against the water concentration. The resulting line was then taken back beyond the origin to give the concentration of residual water (or, strictly, 'water equivalent') in the system; both sets of measurements done with different, but geometrically very similar, apparatus, gave closely similar results around $[H_2O] = 10^{-4}$ mol l^{-1}. The method is illustrated in the generalised diagram, Fig. 4.6.

The concentration of impurities which inactivate a catalyst, C, usually becomes obvious from a plot of the rate constant of the catalysed reaction against [C], such as is shown in Fig. 4.7. As the purification of the solvent, the monomer, and the hardware progresses, not only does the intercept diminish, but the slope of the line (usually) increases. It has been shown that if the dependence of k_1 on monomer concentration, [M], at [C] = constant and on [C] at [M] = constant is available, the [Imp] originating from the solvent and the [Imp] originating from the monomer can be measured separately (Holdcroft and Plesch, 1984).

4.6. What is so special about drying?

Drying, the removal of water, is a special problem because water is a unique compound and there is a lot of it around. The features which together make water unique include: a high dipole moment and polarisability, its amphoteric character, its ability to act as donor and acceptor in hydrogen bond formation, and its smallness.

Much of what needs to be done about the techniques of removing water from glass apparatus and from solvents and reagents has been set out above in a more general context. It remains to point out a few pecularities which characterise the drying process rather than any other purification procedure and to discuss some drying agents.

4.6.1. The drying process

The principal feature which distinguishes drying from other purification processes is this: When one is removing, say, aromatics from cyclohexane, one knows that their quantity is finite and that they are entirely contained within the cyclohexane (we will ignore the fact that because of their greater polarisability the aromatics will be adsorbed on to the water-covered glass surface of the vessel in preference to the alkane). With regard to water the situation is different, because the water in solution in a liquid is in equilibrium with that adsorbed on the surface of the vessel not only where the liquid is in contact with it, but on all other parts of the glassware. From the glass not in contact with the liquid the water can reach the liquid by migration over the surface of the glass and by evaporation from it and condensation into the liquid. Even if most of the adsorbed water has been removed from a glass surface under vacuum, the water layers will be rebuilt as soon as the glass comes in contact with the atmosphere; that is why it is so useful to make the glass hydrophobic by exposing it to trimethyl-chlorosilane (Armstrong, Katovic and Eatham, 1971) because in this way the amount adsorbed can be kept small. Therefore, although strictly the amount of water which needs to be removed from any batch of reagent in given circumstances is, of course, not infinite, it is probably indeterminate and usually greater than that which was merely dissolved.

4.6.2. Some drying agents
Earlier in this chapter we have indicated how sodium and phosphorus pentoxide, certainly two of the best drying agents, should be handled and used. Here we will concentrate on a few other reagents which are used only rarely for purposes other than drying, and which are especially suitable for use in high vacuum systems. In terms of ultimate efficiency they are all very similar and the choice depends on the nature of the compound to be dried, convenience of operation, difficulties of disposing of spent reagents, etc. A comparative study of drying reagents has been published by Burfield and Smithers (1978).

There is little doubt that the 'ultimate weapon' against water is a *molten alkali metal*, which in practice means *sodium above its melting point*, usually in a refluxing solvent. If one needs to operate at ambient temperature and/or to store a compound over a really aggressive drying agent, then the liquid eutectic *mixture of Na + K* is the first choice. A glass-covered magnetic stirrer must be enclosed with the mixture to break up the crust which forms on the liquid.

Evidently, the types of compound compatible with this mixture are not many, but it has been used successfully for various 1,3-dioxacycloalkanes and hydrocarbons. It *must not* be used with halogen compounds although chloro- and bromo-alkanes can be dried safely over sodium films, provided that they are initially at least 'Grignard dry' and free of oxygen. In the author's laboratory specimens of CH_2Cl_2 and MeBr which had been over sodium films under vacuum for several years were analysed and found to contain only traces of the Wurtz-condensation products.

A less aggressive, but almost equally effective, drying agent is *calcium hydride*, provided that it is used correctly. The advantage of CaH_2, in contrast to many drying agents, is that the products of its reaction with water do not adhere to it, but form a scale which falls off the lumps of CaH_2 if it is stirred. Therefore the proper use of this compound requires that one should be able to stir the container effectively; since the reaction with water produces hydrogen, it is also essential that during the early stages when a fresh sample of liquid is mixed with CaH_2, the hydrogen should be pumped off carefully.

Before a batch of CaH_2 is used, a small sample should be mixed with a few drops of water and the evolved gases smelt cautiously (or analysed by GLC). If the nose can sense phosphine or the GLC shows up acetylene, the batch should be returned to the manufacturer. Although in theory both gases would be pumped off during the reevacuation of the liquid compound to remove the hydrogen, enough might remain to interfere with catalytic reactions.

Because of the evolution of hydrogen, it is not advisable to use CaH_2 for the drying of any compound, such as one containing nitro or keto groups, whose reduction would produce compounds which could interfere with the reactions to be studied.

The possibility of an attack by the drying agent on the compound to be dried must always be considered carefully and if necessary examined by small-scale tests, as has been mentioned before. Most notable for such unwanted aggression amongst the popular drying agents is the *molecular sieve*. Various grades of molecular sieve have different activities and specialist publications should be consulted for details. The main reactions which must be kept in mind are that they can catalyse the dehydrochlorination of alkyl chlorides and the condensation of ketones, they can dehydrate alcohols and can form exposive compounds from nitroalkanes (see Section 4.6.3.).

The least spectacular and therefore perhaps least appreciated drying agents are the humble *calcium sulphate* and *sodium sulphate*; they are not widely known because although their drying power is very great, their capacity is modest and for that reason they are not much used by organic chemists. Both compounds should be dehydrated very thoroughly before introduction into a high vacuum system and they can be dried further, or indeed regenerated after use, by being heated under vacuum to well over 100 °C and preferably about 200 °C. They must not be heated so intensely that SO_2 is driven off, because they then become alkaline (CaO or Na_2O) and they start to sinter so that their specific surface is reduced drastically. There are few compounds which cannot be given their ultimate exsiccation by these two compounds.

4.6.3. The special hazards of drying

Scattered throughout the literature one can find numerous warnings which amount to 'Drying may endanger your health'. Therefore it is somewhat astonishing that L. Bretherick's *Handbook of Reactive Chemical Hazards* (3rd edition, Butterworth, London, 1985) contains no special discussion of the hazards of drying. The present short section is not intended to be comprehensive, so if a particular combination of chemical type and drying agent is *not* mentioned, *do not* assume that it is safe! If a new combination of chemical type and drying agent seems to offer some special advantages, and neither chemical common sense nor a search of the literature provides a contra-indication, it is useful to set up a small-scale trial and to monitor by GLC whether the drying agent is doing anything to the chemical to be dried, and if so, what.

Some years ago, $MgClO_4$ was favoured as a drying agent; however, there is now no justification whatever for using it. Because of the explosion danger of perchlorates one should never use any perchlorate for any job that can be done equally well by another type of compound. As for drying these compounds, neither organic nor inorganic perchlorates should be warmed, much less heated, and they must never be ground or comminuted in any other way. The most effective way of drying perchlorates, e.g. for use as the base-electrolyte in electrochemical work, is to pump off on a vacuum line the original solvent from which the salt was precipitated, then distil onto it a solvent such as methylene chloride, and pump that off, so that the residual water co-distils, and repeat that procedure until some suitable test shows the salt to be adequately dry.

Molecular sieves of various grades are very favourite drying agents, but they must not be used indiscriminately. Nitroalkanes must not be dried with them (Bretherick, 1979), and there are reports of the formation of new (and presumably unwanted) compounds under the influence of molecular sieves from 1,1,1-trichloroethane, acetone, and methyl t-butyl ether (Perrin, Armarego and Perrin, 1983).

The attractions of a drying agent which forms a homogeneous mixture with the substance to be dried, e.g. triethyl aluminium or dibutyl magnesium with hydrocarbons and some other compounds, are obvious; the former can be used with methyl methacrylate, the latter with styrene and with dienes. However, it is questionable whether the difficulty of separating the dried compound completely from unused drying agent and the fire-hazard associated with many metal alkyls make the effort worth while, except in some special cases.

One of the alleged hazards of drying, made much of in the 1920s and 1930s, can now be ignored safely. The efforts of one H. B. Baker purported to establish that intensive drying could alter the physical properties, in particular the vapour pressure, of certain liquids. The whole episode has been succinctly reviewed (Farrar, 1963), and the experience of many skillful workers over a further 25 years with systems far drier than Baker ever achieved, enables one to confirm Farrar's view that this was truly a 'mare's nest'.

References

V. C. Armstrong, Z. Katovic and A. M. Eastham, *Canadian J. Chem.*, **49**, 21 (1971).

R. H. Biddulph, P. H. Plesch and P. P. Rutherford, *J. Chem. Soc.*, 275 (1965).

L. Bretherick, *Chem. and Ind.*, 532 (1979).

B. R. Burfield and R. H. Smithers, *J. Org. Chem.*, **43**, 3966 (1978).

J. F. Coetzee, Ed. *Recommended Methods for Purification of Solvents and Tests for Impurities*, Pergamon Books Ltd., Oxford, 1982.

B. Dubois, Thesis, Lille, 1982.

A. M. Eastham, *J. Amer. Chem. Soc.*, **78**, 6040 (1956).

A. M. Eastham and J. M. Clayton, *J. Amer. Chem. Soc.*, **79**, 5368 (1957).

A. G. Evans and G. W. Meadows, *Trans. Faraday Soc.*, **46**, 327 (1950).

F. Fairbrother, N. Scott and H. Prophet, *J. Chem. Soc.*, 1164 (1956).

W. V. Farrar, *Proc. Chem. Soc.*, 125 (1963).

A. Gandini and A. Martinez, *Makromol. Chem., Macromol. Symp.* **13/14**, 211 (1987).

D. W. Grattan and P. H. Plesch, *J. Chem. Soc. Dalton Trans.*, 1734 (1977).

E. A. Hamid, M. Nowakowska and P. H. Plesch, *Makromol. Chem.*, **132**, 1 (1970).

G. E. Holdcroft and P. H. Plesch, *Makromol. Chem.*, **185**, 27 (1984).

Kabir-ud-Din and P. H. Plesch, *J. Electroanal. Chem.*, **93**, 29 (1978).

KPG Keele Polymer Group, 1951–85, traditional, unpublished devices and procedures.

W. R. Longworth, P. H. Plesch and P. P. Rutherford, *International Conference on Co-ordination Chemistry, Chemical Society Special Publ.* No. 113, p. 115, 1959.

W. R. Longworth, C. J. Panton and P. H. Plesch, *J. Chem. Soc.*, 5579 (1965).

B. J. McArdle and J. N. Sherwood, *Chem. and Ind.*, 268 (1985). The authors have issued a 'corrigendum' to the effect that numbers 18–31 in the text correspond to nos. 17–30 in the list of references, and that Reference 16 relates to spinning band fractional distillation.

J. M. Moulis, J. Collomb, A. Gandini and H. Cheradame, *27th IUPAC International Symposium on Macromolecules*, Ed. H. Benoit, IREG, Strasbourg, p. 252, (1981).

T. Natsuume, Y. Akana, K. Tanabe, M. Fujimatsu, M. Shimizu, Y. Shirota, H. Hirata, S. Kusabayashi and H. Mikawa, *Chem. Comm.*, 189 (1969).

T. Natsuume, M. Nishimura, M. Fujimatsu, M. Shimizu, Y. Shirota, H. Hirata, S. Kusabayashi and H. Mikawa. *Polymer J.*, **1**, 181 (1970).

O. Nuyken, S. Kipnich and S. D. Pask, *GIT Fachz.f.d. Laboratorium*, **25**, 461 (1981).

D. D. Perrin, W. L. F. Armarego and D. R. Perrin, *Purification of Laboratory Chemicals*, 2nd Edition, Pergamon Press, Oxford, 1983.

P. H. Plesch, Chapter 18 in *The Chemistry of Cationic Polymerization*, Ed. P. H. Plesch, Pergamon Press, Oxford, 1963.

P. H. Plesch, *J. Macromol. Sci.-Chem.*, **A6**, 980 (1972).

K. Ueno, Ff. Williams, K. Hayashi and S. Okamura, *Trans. Faraday Soc.*, **63**, 1478 (1967).

W. W. Wendlandt, *Thermal Analysis*, 3rd Edition, J. Wiley, New York, Chapter 6, 1986.

5 Chemical operations with high vacuum systems

5.1. Introduction

In this chapter there is a selection of examples of chemical operations done with the help of a high vacuum system. It must be clear to the reader that many chemical operations have already been described in the earlier parts of this book, for example the various purification procedures assembled in Chapter 4. With regard to measurements, several descriptions of how to proceed will be found in Chapter 3, linked to the description of the various instruments.

Evidently, the number of chemical operations which could have been selected for this chapter is vast, and therefore it is appropriate to indicate the points of view which governed the choices. The first consideration was to include especially useful but unpublished procedures which are 'traditional' in some laboratories, or which are described only in theses, or which have only been described sketchily in the literature.

The second consideration was to bring procedures from laboratories other than the author's, and also those which do not deal with polymerisations. The reason for this is that the author is very conscious of the two biases which colour much of this book, and he wants to show as impressively as he can the great utility of h.v.t. to all kinds of chemists. The choices made from the remaining mass of examples are largely a matter of chance and there is little doubt that any knowledgeable colleague would have made a different selection. One aim has been to bring contributions from as many research groups as was possible, so as to present a wide variety of working styles.

Last, the very different treatments given to the examples needs explanation. Roughly speaking, the more inaccessible the original publication, the more detail has been given, with unpublished procedures being described in full.

5.2. Syntheses

5.2.1. Sodium ethoxide by the cascade method

When sodium ethoxide is made from ethanol, dried by distillation from sodium in vacuum, and vacuum-distilled sodium, the product is always yellow. The author believes that he heard the method of obtaining the pure, colourless product in M. Szwarc's laboratory and that it is unpublished; the following description is adapted from a Keele thesis.

Sodium was purified in the apparatus shown in Fig. 5.1(a), thus: Ca. 12 g of sodium, cleaned as described in Section 4.4.3.1, was introduced into a tube A, which was then sealed to the apparatus at a; this was evacuated for 8 h without heating the sodium, and then sealed off from the vacuum line at e. Tube A was then heated gently and the molten sodium poured swiftly into B, leaving its skin stuck to the tube, which was sealed off at b. This process was repeated by pouring the sodium successively into C and D and finally collecting the silvery metal in E, the sections being sealed off successively at c, d, and f. It is not known why this method of purification is more effective than distillation. The flask E was reattached to the vacuum line via the break-seal g (Fig. 5.1(b)) and then 300 ml of purified ethanol was distilled into E from the container F.

The reaction between sodium and ethanol was exceedingly slow under high vacuum. As sodium ethoxide was precipitated on the sodium, inhibiting the reaction, it was necessary to stir and heat the reaction mixture. Despite this, it took three days for the reaction to go to completion. The hydrogen evolved was pumped off from time to time. When all the sodium had dissolved, the excess of ethanol was pumped off, which left white crystals of sodium ethoxide in E. These were dissolved in diethyl ether, distilled into E from the container G, then E was sealed off from the line at h and fused onto the tipping device (Fig. 5.1(c)) via break-seal i; the solution was colourless. After evacuation, the tipping device was sealed off from the line at j, seal i was broken, and the solution poured through the sintered filters S_1 and S_2 into reservoir H. Then the phials P were filled by judicious pouring, and the excess of the sodium ethoxide solution was transferred to J and sealed off for further use. Finally, the phials P were equilibrated as described in Section 3.1.2 and sealed off.

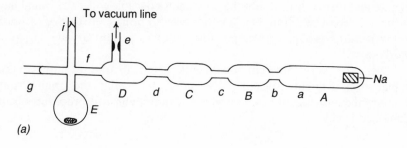

(a)

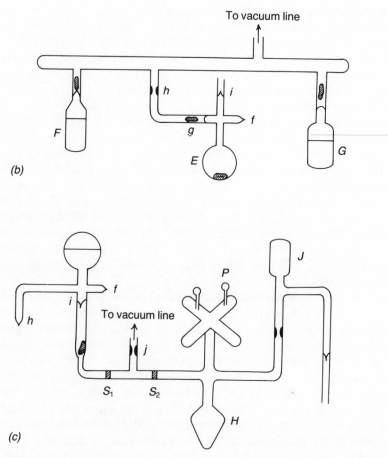

(b)

(c)

Fig. 5.1. The cascade purifier-reactor for the preparation of sodium ethoxide.

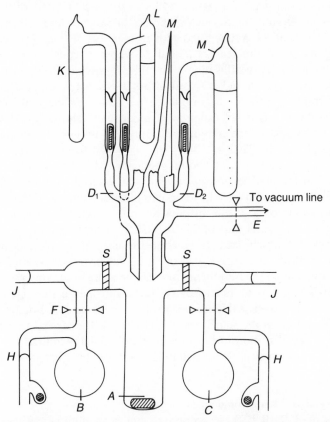

Fig. 5.2. Apparatus for the preparation and purification of $BuOTiCl_3$.
Container K holds the $TiCl_4$, L holds $(BuO)_4Ti$, and at the three tubes marked
M ampoules with CH_2Cl_2, identical to that shown, are attached.

5.2.2. n-Butoxytitanium trichloride

The title compound was required as an initiator for cationic polymerisations.
The account of its synthesis (Vairon and Sigwalt, 1971) is interesting for
several reasons: The chemical reaction (5.1) was selected in preference to an
alternative $(TiCl_4 + n\text{-}BuOH)$ because there are no by-products:

$$3\ TiCl_4 + (BuO)_4Ti \rightarrow 4\ BuOTiCl_3 \qquad (5.1)$$

Both reagents were purified (which includes drying) in vacuum systems and,
the synthesis having been done in the same way, further drying was deemed
unnecessary, and the product was purified only by crystallisation. The
synthesis also provides one of the best examples of the style of apparatus and
procedure used by Sigwalt's school. The purification of the $TiCl_4$ and the
$(BuO)_4Ti$ are described in exhaustive detail, but here we will give only a
slightly abbreviated version of the actual synthesis. See Fig. 5.2.

The apparatus consists of a central cylindrical vessel A fitted with two trifurcated spouts D_1 and D_2 to which ampoules of reagents and solvent are sealed. On each side of A is a tube fitted with a sintered glass filter S (grade 3) giving access to the dump-vessels B and C whose purpose is to store supernatant liquid from crystallisations. The tube E provides the connection to the high vacuum system and is sealed off after exhaustive evacuation of the whole assembly before the reaction is carried out. After the walls of the assembly have been rinsed, as usual, with methylene chloride dried over sodium, the break-seal to the ampoule of $(BuO)_4Ti$ is broken and this is poured into A. Then the $TiCl_4$ is added similarly, and the reaction sets in immediately, the mixture being stirred magnetically. After 1 h the mixture is almost solid, and the first portion of solvent is introduced from one of the ampoules which results in a very fast and complete dissolution of the solid. The pale yellow solution is left in the dark at $-30\,^{\circ}C$ which induces a slow crystallisation of the greater part of the $BuOTiCl_3$. The supernatant liquid is filtered into vessel B which needs to be cooled to effect a complete transfer of the solvent, and which is then sealed off at F. The break-seal H serves to seal B onto a vacuum line so that further use can be made of its contents. The crystallisation is repeated by means of vessel C, and then a third time by means of a further receiver sealed on at one of the break-seals J. Finally, the $BuOTiCl_3$ is dissolved in the fourth lot of CH_2Cl_2 and transferred to a new rig for dilution and distribution into the breakable phials required for the polymerisation experiments, by a technique described by Cheradame and Sigwalt (1970).

5.2.3. Tertiary silyl lithium compounds

The synthesis of $R^1R^{11}R^{111}$ SiLi compounds under vacuum (Evans, Jones and Rees, 1967) provides a contrast to the previous example, being very simple, and is a useful instance of the adaptation of a 'conventional' method (Gilman and Lichtenwalter, 1958) to h.v.t. The objects of the syntheses were $Me_2PhSiLi$, $MePh_2SiLi$ and Ph_3SiLi whose reaction with 1,1-diphenylethylene was to be studied. The preparation of all three organo-Li compounds was very similar, and the reaction is shown in Equation (5.2):

$$(MePh_2Si)_2 + 2\ Li \rightarrow 2\ MePh_2SiLi \tag{5.2}$$

The apparatus is shown in Fig. 5.3.

The disilane was introduced through the side arm into vessel C and the whole apparatus evacuated with heating. The disilane was then sublimed into D which was sealed off at B. A known quantity of THF was distilled into D, from the THF supply H, after which the rig was sealed off at E and F. Previous to this, a film of Li had been deposited in A (Garst and Zabotny, 1965). The solution of the disilane in THF was run into A which was then sealed off at G. The reaction was left to continue at $0\,^{\circ}C$ for one week, and the final, filtered solution was stored in a sealed ampoule at $-25\,^{\circ}C$.

To vacuum line

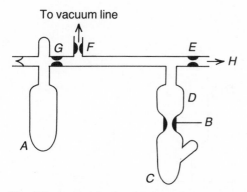

Fig. 5.3. Apparatus for the preparation of organosilyl lithium compounds.

The omission of details such as the making of the Li film, the filtering, or the exact manner of using the reagent, shows that Evans *et al.* assume a certain familiarity with h.v.t. on the part of their readers, in marked contrast to Vairon and Sigwalt.

5.2.4. 1,3-Dioxolenium salts

When a strong electrophile, such as a triphenyl methylium ion, abstracts a hydride ion from the 2-position of 1,3-dioxolan, the product is the 1,3-dioxolan-2-ylium or (for short) dioxolenium ion as shown in Equation (5.3):

$$Ph_3C^+A^- + CH_2\!\!\begin{array}{c}O\\ \diagdown\\ O\end{array}\!\!\rightarrow Ph_3CH + \begin{array}{c}O\\ \diagup\\ O\end{array}\!\!\!\!\overset{+}{\diagup}\!\!CH\ A^- \qquad (5.3)$$

Salts of this type have been used by Penczek's school as initiators for the polymerisation of heterocyclic monomers, and their preparation (Stolarczyk, Kubisa and Penczek, 1977) is selected here as being typical for such syntheses and a good example of the Lodz school's manner of working. (See also p. 93)

The apparatus is shown in Fig. 5.4. Commercial $Ph_3C^+SbF_6^-$ (3 g) was dissolved in CH_2Cl_2, precipitated in CCl_4 and introduced into tube A of the apparatus which was then evacuated through C_2. Then 40 ml of CH_2Cl_2 stored over CaH_2, and subsequently 1.5 g of dioxolan, stored over liquid $Na + K$ alloy, was distilled into A. After a few minutes at 0 °C the dioxolenium salt started to crystallise out and after 30 min the yellow colour of the Ph_3C^+ ion had disappeared. After closing the PTFE tap T_1, the apparatus was detached from the vacuum line at C_2 and attached through cone C_1 to socket S_1 of the adaptor B which was attached to the vacuum line through cone C_3 in an appropriately inclined attitude. The solution passed through the filter F into B, leaving the salt in A. This procedure was repeated with fresh portions of CH_2Cl_2 until the salt was colourless. Then a weighed phial D was attached to cone C_2 and evacuated, with the apparatus inverted, and some of

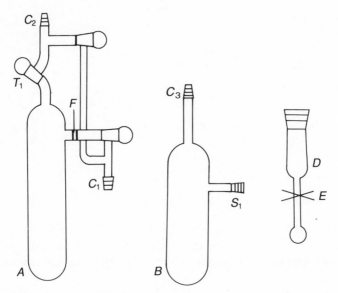

Fig. 5.4. Apparatus for the synthesis, purification and distribution into phials of 1,3-dioxolane-2-ylium salts.

the salt transferred from A to D through T_1, and the phial itself was sealed off at E. This procedure was repeated until all the required phials had been filled.

It is obvious that this technique is less rigorous than those described earlier and that it could have been made more efficient by the use of a tipping device. Nonetheless, it seems to have been adequate for the intended work.

5.2.5. Electrochemical preparation of stannic chloride

Evans and Lewis (1957) have described an ingenious method of preparing $SnCl_4$ electrochemically in a high vacuum system from $SnCl_2$ which, being crystalline at ambient temperature, is easier to purify than the tetrachloride. The apparatus is shown in Fig. 5.5.

Anhydrous 'AnalaR' stannous chloride was distilled from A into the electrolytic cell, C, under a high vacuum after the whole apparatus had been baked out. Vessel A was sealed off at X, the stannous chloride fused by the heater H, and a 12 V DC potential applied across the terminals T_1 and T_2; T_1 led to a gas-carbon rod serving as anode, T_2 to a Pt wire cathode. Bubbles of stannic chloride vapour rose from the anode. The first runnings were pumped out and then the system was sealed off at Y. Electrolysis was continued until sufficient stannic chloride had been condensed in the sample holder, S, which was removed from the system by being sealed off at Z.

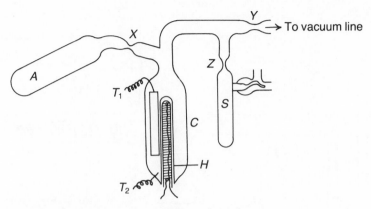

Fig. 5.5. Electrolytic cell for the preparation of $SnCl_4$.

5.2.6. Aluminium chloride

The synthesis of $AlCl_3$ can be done in several ways: from the elements, from $Al + HCl$, and by displacement of Ag from molten AgCl:

$$3AgCl + Al \rightarrow AlCl_3 + Ag$$

The first method has little to recommend it, because of the difficulty of purifying chlorine. The second method was favoured by Fairbrother, Scott and Prophet (1956) for their studies on the solubility of $AlCl_3$ in various hydrocarbons. Dry HCl was passed over Al wire kept at 400–500 °C, the product was sublimed four times in a high vacuum system and sealed into a series of break-seal ampoules. Although it is not stated explicitly, this procedure involves the (slight) difficulty of doing the actual synthesis more or less at atmospheric pressure, and then changing to vacuum operation.

Probably the easiest and most efficient method is the third (Wallace and Willard, 1950) which involves dropping Al wire into molten AgCl. The object was to label the $AlCl_3$ with ^{36}Cl, so the AgCl was made from $H^{36}Cl$ (aq.) and $AgNO_3$. The original method was used in the author's laboratory several times, and the design of apparatus shown here (Fig. 5.6) and the procedure are based on the last version (KPG).

Spirals of 'four nines' Al wire were rinsed in dilute HCl, then in ethanol and introduced into tube A infront of the glass-enclosed iron core B. The well-dried AgCl, synthesised from AnalaR $AgNO_3$ and HCl, was introduced into the Kon flask C and the whole rig was then evacuated whilst the AgCl was warmed gently. Then the apparatus was sealed off from the vacuum line at D, the AgCl was melted by stronger heating, and the first spiral of Al wire was pushed magnetically to drop into the melt. The reaction started rapidly and became quite violent, and the purpose of the Z-bend is to prevent

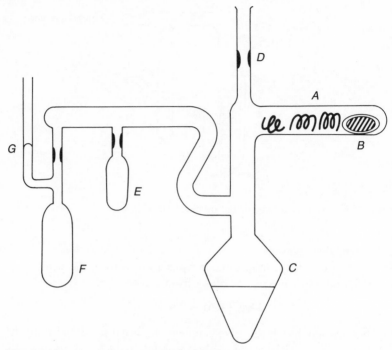

Fig. 5.6. Apparatus for the synthesis of AlCl$_3$.

particles of the reaction mixture being carried into the main line. The first portion of AlCl$_3$ which sublimed out was sealed off in the ampoule E and discarded as it was always coloured (probably from traces of heavy metals extracted from the glass). The remainder of the AlCl$_3$, generated by further additions of Al, was sublimed by means of a brush-flame into F, which was then sealed off. Subsequently it was fused via the break-seal G onto a tipping device and the AlCl$_3$ sublimed into phials.

5.2.7. Complex salts

Aroyl and related salts (ArCO$^+$, AlkCO$^+$) and stable carbenium ion salts (e.g. Ar$_3$C$^+$) of complex anions MtX$_{n+1}^-$ can be synthesised essentially by two routes, silver salt double decomposition, and direct combination:

$$RCOCl + AgMtF_{n+1}^- \rightarrow AgCl\downarrow + RCO^+MtF_{n+1}^-$$

$$RCOF + MtF_n \rightarrow RCO^+MtF_{n+1}^-$$

The first method has the advantages that the reagents are cheaper, more readily available and easier to handle, but the disadvantage that the product must be isolated from the unwanted AgCl. The synthetic procedure and the apparatus are correspondingly complicated.

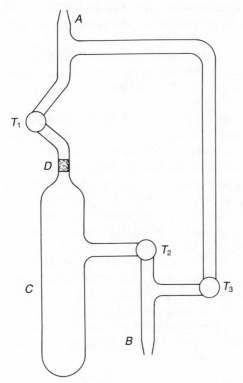

Fig. 5.7. Reactor for preparing complex salts.

The direct combination method is very much simpler, faster and cleaner, since any excess of either reagent can be simply pumped away. However, volatile fluorides (BF_3, PF_5, SbF_5) are extremely aggressive reagents. The required apparatus is shown in Fig. 5.7, in which all the taps are of PTFE; the synthesis of the SbF_6^- salt will be described, as SbF_5 is more difficult to manipulate than most other volatile fluorides (Nuyken, Kipnich and Pask, 1981). In the following account the words 'attached' and 'detached' can signify operations involving sealing on or off, or the use of cone and socket joints. If these are used, the cones should be on the attachment, the socket on the vacuum line, and the joint must be sealed with a fluorinated material or a metal (Au) wire; on no account must a silicone or hydrocarbon lubricant be used.

The rig is attached to the vacuum line at A, and the ampoules containing the solvent (SO_2), precipitant (Freon 113) and the reagents are attached at B; either these ampoules contain the exact amounts required, or the subsidiary line to which they are attached must also comprise a hanging burette. Since the normal boiling point of SO_2 is -10 °C and its vapour pressure at ambient temperature $ca.$ 3 atm., the appropriate precautions must be taken, i.e. the apparatus must not contain any thin-walled parts and the SO_2 and solutions in it should be kept at not more than 0 °C, and preferably at a lower temperature.

After thorough evacuation and removal of surface water by brush-flaming or silylation, the SbF_5 is distilled into C, then the SO_2, and finally the aroyl fluoride. The reaction should be done at about 0 °C and it is over rapidly.

The SbF_5 is a viscous, colourless, liquid of high boiling point which distils but slowly. Therefore it is advisable to work with an excess (*ca.* 10%) of the organic fluoride which can be pumped away after completion of the synthesis much more rapidly and efficiently than any residual SbF_5, which in any case would mess up the vacuum line (any specks of grease or other organic matter would turn black), and is awkward to remove from the cold trap. About half of the SO_2 is pumped off (or distilled into a dump vessel), then the Freon 113 is distilled in, whereupon the product is precipitated as fine, colourless, crystals. With all taps closed, the subsidiary rig is detached at B, the apparatus is detached at A and reattached to the line at B, having been turned through 180°. A receiver (Kon flask) is attached at A and evacuated via T_3. By cooling this receiver with T_1 open and T_2 closed, the liquid in C is drawn off through the frit D which retains the salt. This is washed several times with fresh Freon from a supply on the main line, and then pumped dry until it is a free-running powder. As such it is easy to transfer to a tipping device through B and thence to phials.

5.3. Measurements

In Chapter 3 we have described many instruments for making a variety of measurements on reaction mixtures under high vacuum. As their method of use is quite clear and some details are given for many, we give here only a few examples to illustrate some particularly interesting and useful points.

5.3.1. Electrical conductivity and UV measurements
The electrical conductivity of solutions is measured for at least three distinct purposes:

(1) To investigate ionic equilibria so as to obtain information about the interactions of ions with each other, with non-ionic solutes and with the solvent.

(2) To measure the rates of chemical reactions in which the ionic population changes, i.e. reactions during which the nature and/or number of ions changes.

(3) To monitor what is going on in a solution without necessarily wanting to interpret the changes quantitatively. This is akin to following changes of temperature or of colour.

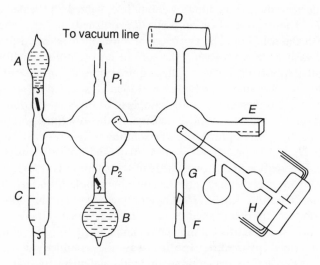

Fig. 5.8. Combined device for determining the conductivity and the UV absorption spectrum of living polystyrene.

As many of the electrochemically most interesting systems are very sensitive to water, either because it reacts with one or more of the ions under study or because it solvates ions more strongly than the solvent concerned, there has always been a strong incentive for electrochemists to use h.v.t. It is therefore not surprising that an immense number of vacuum conductivity cells have been described; and as most electrochemical experiments involve measurements of conductivity as a function of the concentration of one or more species, much ingenuity has been expended on diluting or concentrating or titrating a solution under vacuum.

The example selected here shows the techniques developed by Szwarc's group (Bhattacharyya, Lee, Smid and Szwarc, 1965). They eschewed completely the use of taps and in each experiment the electrical conductivity and the optical absorption of the solution were measured; this actually involves the same combination of devices as the Pask–Nuyken and Holdcroft–Plesch reactors described in Section 3.2.2. The description of the procedure is taken almost verbatim from the original publication, whose conciseness cannot be bettered.

The conductivities of living polystyrene in THF were determined in the apparatus shown in Fig. 5.8. (In this context 'living polystyrene' means a chain ending in $—CHPh^-Mt^+$.) It consists of a sealed conductivity cell, H, and three optical cells, D, E and F, having optical path lengths of 10, 1 and 0.2 cm, respectively, the last one being equipped with a 0.19 cm spacer. Flask B contains a living polymer solution to be used for purging the apparatus and destroying all the residual impurities adsorbed on the walls. Ampoule A contains a fairly concentrated solution of the polymer to be investigated.

The apparatus is evacuated on a high vacuum line, flamed and sealed off at constriction P_1. The break-seal on ampoule B is crushed, and the whole unit is rinsed with the solution. The rinsing solution is then returned to B, and the unit is 'washed' by condensing the solvent on the walls which are chilled in the usual way from outside. The required amount of solvent is distilled from B to a graduated cylinder, C, and then B is sealed off at P_2.

The break-seal on A is then crushed, the sample to be investigated diluted with the solvent in C, and the optical density of the resulting solution determined in the appropriate optical cell. The solution is then transferred to conductivity cell H, and its resistance is measured. The optical density is redetermined, and thereafter about two thirds of the solution is transferred to C. The solvent from C is distilled into the chilled ampoule G and used to dilute the residual solutions left in H. The conductivity and the optical density of this solution are determined as described previously; thereafter, two thirds is again transferred to C, and the remaining one third is diluted by the above-described procedure. In this way the conductivities are determined for decreasing concentrations of living polymer, so that the molar conductivity Λ can be calculated as a function of [living polymer] down to about 10^{-5} M.

The described method has several advantages: (a) The products derived from impurities are removed from the system together with the purging solution, and, therefore, they do not contribute to the measured conductivity. Such a contribution might be appreciable at extremely low concentrations of living polymer. (b) No impurities are introduced on dilution, since the same solvent is used over and over again. (c) The destruction or 'isomerisation' of living polymers, which sometimes occurs in a highly dilute solution, is minimised since one proceeds from a more concentrated to a more dilute solution.

5.3.2. Combined UV and ESR spectroscopy.
In the course of an extensive study of the mode of action of anionic catalysts, A. G. Evans used exclusively h.v.t. His investigation of the electron-transfer reactions involving 1,1,3,3-tetraphenylbutene-1, tetraphenylethylene and sodium naphthalide is of particular interest here because all the reaction mixtures were prepared by h.v.t. and both UV and ESR spectra were measured (J. E. Bennett et al., 1963). The paper contains full experimental details.

5.3.3. Phase diagrams
One of the most unambiguous and simple methods of investigating the association of molecules, as in charge-transfer complexes, solvates, etc. is by means of the freezing point or the vapour pressure of the mixture as a function of its composition. These 'classical' methods have been displaced to some extent by spectroscopic methods, which, however, fail if one is dealing

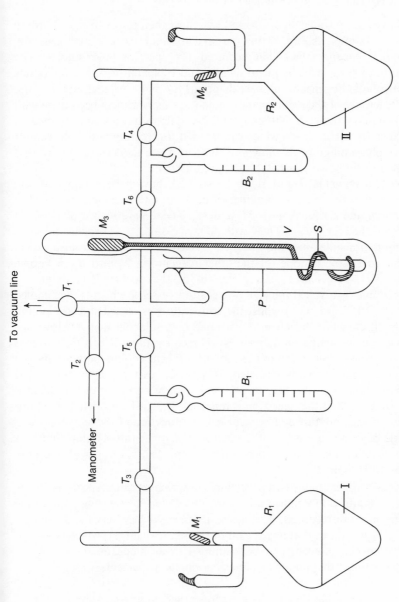

Fig. 5.9. Apparatus for simultaneous determination of freezing point and vapour pressure curves of mixtures (phase diagrams).

with fairly concentrated solution, and one must then resort to the recording of phase-relations; the apparatus required for such work is also generally very much cheaper than that required for spectroscopic studies. The formation of complexes by metal halides and by metal-organic compounds has been studied by means of phase diagrams for many decades and, because of the reactivity of the complexands, much of this work has been done with high vacuum systems, so that many different devices have been published. A very generally useful set-up which can be used for freezing point and vapour pressure phase diagrams is shown in Fig. 5.9; it is a modern version of that described by Longworth, Plesch and Rigbi (1958).

The two reagents, I and II, to be studied, having been prepared and purified under vacuum, are contained in flasks R_1 and R_2 which are sealed to the vacuum line via break-seals M_1 and M_2. The apparatus is evacuated via T_1 as far as the break-seals. Then, with all taps closed, M_1 is broken and some of its content distilled into burette B_1, T_3 is closed and B_1 is thermostatted in ice. Then the same is done for the other reagent via T_4 into B_2. A known volume of I, sufficient to cover the tip of the pocket P containing a thermocouple, is distilled into the observation vessel V which is then cooled slowly (5°–10 °C per min.) whilst the temperature is recorded as a function of time; all the while the stirrer S, operated by a solenoid with an electronic 'flip-flop' device acting on magnet M_3, is reciprocating. When the stirrer is stopped by the solidification of the content of V, the cooling bath is removed and the warming curve is recorded. When the freezing point (strictly, the triple point) of pure I has been established, a known amount of II is distilled into V from B_2, and the operation is repeated. If T_2 is kept open, the vapour pressure of each mixture can be recorded as a function of the temperature at the same time. When V is full, its contents are distilled into a dump container on the vacuum line, and the procedure is repeated, starting with the pure compound II from B_2.

Several workers have used procedures in which measured portions of the reagents contained in weighed phials are mixed, and the freezing point of the resulting mixture recorded. The method described here has the advantage over such procedures that any region of the phase diagram, which turns out to be of special interest, can be examined in as much detail as may be required without the labour of preparing phials of reagents.

5.3.4. Polymerisation kinetics by gel permeation chromatography (GPC)

It may happen that the properties of a monomer and its oligomers and polymers are so similar that they cannot be separated by precipitation; that in the appropriate solvents the reaction mixture gels, so that dilatometry cannot be used; that it becomes so opaque that neither refractive index nor optical rotation can be determined; and that the reaction is too slow for normal reaction calorimetry. This situation was met when the author attempted to study the polymerisation of trimethyl and tribenzyl laevo-

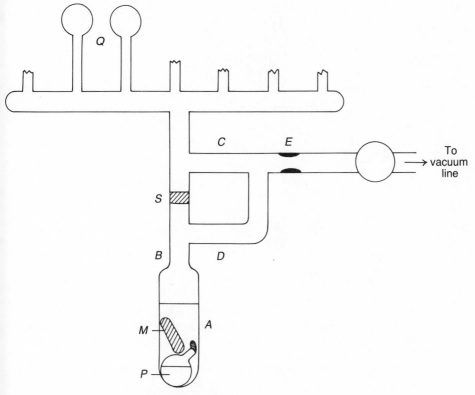

Fig. 5.10. Apparatus for measuring polymerisation kinetics by GPC.

glucosan in methylene chloride and other solvents by cationic initiators. The difficulties encountered explain the paucity of kinetic results in the literature. As the method devised to circumvent these difficulties has not yet been published, but is widely applicable, it is presented here.

The principle is this: A reaction mixture is made up and distributed into breakable phials. These are thermostatted, broken at noted times into a neutralising solution, and the mixture analysed immediately by GPC. This was done in the apparatus shown in Fig. 5.10 as follows.

The ampoule A was charged with the crystalline monomer, a phial P of initiator (e.g. PF_5 solution), and a magnetic breaker M and then sealed at B to the rest of the rig which was then evacuated. Methylene chloride was distilled in from a reservoir on the vacuum line to dissolve the monomer and was then pumped off slowly, and evacuation was continued for 8 h, which process produced an efficient final drying. The purpose of the two pumping ducts C and D is to circumvent the evacuation obstacle presented by the sintered filter S. The required volume of solvent was then distilled into A, the rig sealed off from the line at E, and the ampoule A brought to the right

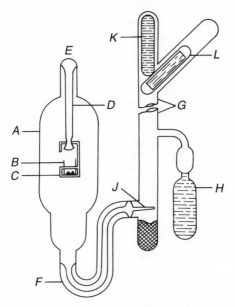

Fig. 5.11. Vacuum reactor for chlorination of metals. *A* Reaction vessel (100 ml); *B* beryllia crucible containing titanium metal; *C* silica cradle; *D* crucible support also serving as evacuation duct, and finally sealed off at the top at *E*; *F* capillary tube; *G* duct for breaker; *H* appendix containing liquid chlorine; *J* fragile capillary tip; *K* weighted glass breaker; *L* glass-coated magnetic retainer.

temperature. Then the phial *P* was broken, the solution mixed well by shaking and distributed into the phials *Q*; these were then immersed in the thermostat and sealed off. Each phial was placed into a boiling tube containing a few millilitres of a neutralising solution, e.g. triethylamine in methylene chloride, and stored in the thermostat. At noted time intervals, each phial was broken by a glass rod into the neutralising solution, so that the polymerising mixture was neutralised almost instantly at the reaction temperature; this is very important for monomers, such as those mentioned here, which participate in a monomer–polymer equilibrium to prevent depolymerisation by warming. A sample of the resulting mixture was injected into a GPC machine, the output from which was fed into a microcomputer and was displayed as the molecular weight distribution curve, and simultaneously the ratio of the amount of residual monomer to that of reaction product, i.e. the yield of the reaction, was calculated. This method has the further advantage that the changes in the molecular weight distribution with time are immediately available.

    It was found unnecessary to remove solvent, initiator and neutralising agent before the GPC analysis. Evidently, this method is applicable to any kind of reaction, provided that it is sufficiently slow, and that a suitable method of analysis can be found.

5.3.5. Precision calorimetry

The determination of enthalpies of reaction for reactions involving reagents or products which are sensitive to any of the components of the atmosphere requires that the reactions be carried out in an inert atmosphere or in a vacuum. P. Gross and co-workers who were concerned with the enthalpies of formation of water-sensitive metal halides such as titanium tetrachloride, devised a vacuum reactor which could be immersed totally in the water bath of a precision calorimeter; it is illustrated in Fig. 5.11. (Gross, Hayman and Levi, 1955). The reaction between the chlorine in H and the titanium in C is initiated by lifting the glass-coated retainer L so that the weighted glass breaker K ruptures the tip J of the capillary F. Improved and modified designs were published later (Gross, Hayman and Levi, 1957a,b).

References

J. E. Bennett, A. G. Evans, J. C. Evans, E. D. Owen and B. J. Tabner, *J. Chem. Soc.*, 3954 (1963).

D. N. Bhattacharyya, C. L. Lee, J. Smid and M. Szwarc, *J. Phys. Chem.*, **69**, 612 (1965).

H. Cheradame and P. Sigwalt, *Bull. Soc. Chim. France*, 843 (1970).

A. G. Evans and J. Lewis, *J. Chem. Soc.*, 2975 (1957).

A. G. Evans, M. Ll. Jones and N. H. Rees, *J. Chem. Soc. (B)*, 961 (1967).

F. Fairbrother, N. Scott and H. Prophet, *J. Chem. Soc.*, 1164 (1956).

J. F. Garst and E. R. Zabotny, *J. Amer. Chem. Soc.*, **87**, 495 (1965).

H. Gilman and G. D. Lichtenwalter, *J. Amer. Chem. Soc.*, **80**, 608 (1958).

P. Gross, C. Hayman and D. L. Levi, *Trans. Faraday Soc.*, **51**, 626 (1955).

P. Gross, C. Hayman and D. L. Levi, *Trans. Faraday Soc.*, **53**, 1285 (1957a).

P. Gross, C. Hayman and D. L. Levi, *Trans. Faraday Soc.*, **53**, 1601 (1957b).

KPG, Keele Polymer Group, unpublished.

W. R. Longworth, P. H. Plesch and M. Rigbi, *J. Chem. Soc.*, 451 (1958).

O. Nuyken, S. Kipnich and S. D. Pask, *GIT Fachz. f.d. Laboratorium*, **25**, 445 (1981).

A. Stolarczyk, P. Kubisa and S. Penczek, *J. Macromol. Sci.-Chem.* **A.11**, 2047 (1977).

J. P. Vairon and P. Sigwalt, *Bull. Soc. Chim. France*, 559 (1971).

C. H. Wallace and J. E. Willard, *J. Amer. Chem. Soc.*, **72**, 5275 (1950).

Appendix Suppliers of vacuum equipment

A.1 General suppliers

Cole-Parmer Instr. Co.,
7425 N. Oak Park Avn., Chicago, Ill. 60648, USA
Edwards High Vacuum,
Manor Royal, Crawley, West Sussex, RH10 2LW, England.
Genevac Ltd.,
Alpha Works, White House Road, Ipswich, IP1 5LU, England.
Leybold-Heraeus,
Bonner Str. 498, D-5000 Koeln, FRG
Veeco Instruments Ltd.,
Abbey Barn Road, High Wycombe, Bucks., HP11 1RW, England.

A.2 Special equipment

A.2.1. Glassware, including PTFE taps
Bibby Science Products Ltd.,
Tilling Drive, Walton Industrial Estate, Stone, Staffs. ST15 0SA,
England. Tel (9782) 812121, Telex 36225, FAX 0785–813748.
J. Young (Scientific Glassware) Ltd.,
11 Colville Road, Acton, London, W3 8BS, England. Tel (01) 992 0891.

A.2.2. Glove boxes
Faircrest Engineering Ltd
4 Union Road, Croydon, CR0 2XX, England. Tel (01) 689 8741.
Tempo Trading Ltd.,
6 Rising Sun Industrial Estate, Blaina, Gwent, NP3 3JW, UK

A.2.3. Metal valves, connectors, etc.
Cajon Company,
9760 Shepard Road, Macedonia, Ohio 44056, USA
Hoke Inc.,
1 Tenakill Park, Cresskill, N.J. 07626, USA.
Nupro Company,
4800 East 345th Street, Willoughby, Ohio 44094, USA.
Swagelok Crawford Fitting Company,
29500 Solon Road, Solon, Ohio 44139, USA
Veeco,
Terminal Drive, Plainview, New York 11803, USA

A.2.4. Microbalances for weighing in vacuum
C.I. Electronics Limited,
Brunel Road, Churchfields, Salisbury, Wilts, England. Tel (0772) 336938,
Telex 47395, FAX 0722–23222.

Name index

Subject index